PRACTICAL MATHEMATICAL OPTIMIZATION

An Introduction to Basic Optimization Theory and Classical and New Gradient-Based Algorithms

Applied Optimization

VOLUME 97

Series Editors:

Panos M. Pardalos
University of Florida, U.S.A.

Donald W. Hearn
University of Florida, U.S.A.

PRACTICAL MATHEMATICAL OPTIMIZATION

An Introduction to Basic Optimization Theory and Classical and New Gradient-Based Algorithms

By

JAN A. SNYMAN
University of Pretoria, Pretoria, South Africa

Springer

Library of Congress Control Number: 2005934835

ISBN-10: 0-387-29824-X e-ISBN: 0-387-24349-6

ISBN-13: 978-0387-24348-1

Printed on acid-free paper.

AMS Subject Classifications: 65K05, 90C30, 90-00

Printed in the United States of America.

9 8 7 6 5 4 3 2 1

springeronline.com

To

Alta

my wife and friend

Contents

Preface

It is intended that this book be used in senior- to graduate-level semester courses in optimization, as offered in mathematics, engineering, computer science and operations research departments. Hopefully this book will also be useful to practising professionals in the workplace.

The contents of the book represent the fundamental optimization material collected and used by the author, over a period of more than twenty years, in teaching Practical Mathematical Optimization to undergraduate as well as graduate engineering and science students at the University of Pretoria. The principal motivation for writing this work has not been the teaching of mathematics per se, but to equip students with the necessary fundamental optimization theory and algorithms, so as to enable them to solve practical problems in their own particular principal fields of interest, be it physics, chemistry, engineering design or business economics. The particular approach adopted here follows from the author's own personal experiences in doing research in solid-state physics and in mechanical engineering design, where he was constantly confronted by problems that can most easily and directly be solved via the judicious use of mathematical optimization techniques. This book is, however, not a collection of case studies restricted to the above-mentioned specialized research areas, but is intended to convey the basic optimization principles and algorithms to a general audience in such a way that, hopefully, the application to their own practical areas of interest will be relatively simple and straightforward.

Many excellent and more comprehensive texts on practical mathematical optimization have of course been written in the past, and I am much indebted to many of these authors for the direct and indirect influence their work has had in the writing of this monograph. In the text I have tried as far as possible to give due recognition to their contributions. Here, however, I wish to single out the excellent and possibly underrated book of D. A. Wismer and R. Chattergy (1978), which served to introduce the topic of nonlinear optimization to me many years ago, and which has more than casually influenced this work.

With so many excellent texts on the topic of mathematical optimization available, the question can justifiably be posed: Why another book and what is different here? Here I believe, for the first time in a relatively brief and introductory work, due attention is paid to certain inhibiting difficulties that can occur when fundamental and classical gradient-based algorithms are applied to real-world problems. Often students, after having mastered the basic theory and algorithms, are disappointed to find that due to real-world complications (such as the presence of noise and discontinuities in the functions, the expense of function evaluations and an excessive large number of variables), the basic algorithms they have been taught are of little value. They then discard, for example, gradient-based algorithms and resort to alternative non-fundamental methods. Here, in Chapter 4 on new gradient-based methods, developed by the author and his co-workers, the above mentioned inhibiting real-world difficulties are discussed, and it is shown how these optimization difficulties may be overcome without totally discarding the fundamental gradient-based approach.

The reader may also find the organisation of the material in this book somewhat novel. The first three chapters present the basic theory, and classical unconstrained and constrained algorithms, in a straightforward manner with almost no formal statement of theorems and presentation of proofs. Theorems are of course of importance, not only for the more mathematically inclined students, but also for practical people interested in constructing and developing new algorithms. Therefore some of the more important fundamental theorems and proofs are presented separately in Chapter 6. Where relevant, these theorems are referred to in the first three chapters. Also, in order to prevent cluttering, the presentation of the basic material in Chapters 1 to 3 is interspersed with very few worked out examples. Instead, a generous number of worked out example problems are presented separately in Chapter 5, in more or less the same order as the presentation of the corresponding theory given in Chapters 1 to 3. The separate presentation of the example problems may also be convenient for students who have to prepare for the inevitable tests and examinations. The instructor may also use these examples as models to easily formulate similar problems as additional exercises for the students, and for test purposes.

Although the emphasis of this work is intentionally almost exclusively on gradient-based methods for non-linear problems, the book will not be complete if only casual reference is made to the simplex method for solving Linear Programming (LP) problems (where of course use is also made of gradient information in the manipulation of the gradient vector $\mathbf{c}$ of the objective function, and the gradient vectors of the constraint functions contained in the matrix $\mathbf{A}$). It was therefore decided to include, as Appendix A, a short introduction to the simplex method for LP problems. This appendix introduces the simplex method along the lines given by Chvatel (1983) in his excellent treatment of the subject.

The author gratefully acknowledges the input and constructive comments of the following colleagues to different parts of this work: Nielen Stander, Albert Groenwold, Ken Craig and Danie de Kock. A special word of thanks goes to Alex Hay. Not only did he significantly contribute to the contents of Chapter 4, but he also helped with the production of most of the figures, and in the final editing of the manuscript. Thanks also to Craig Long who assisted with final corrections and to Alna van der Merwe who typed the first LaTeX draft.

Jan Snyman

Pretoria

Table of notation

$\mathbb{R}^n$	n-dimensional Euclidean (real) space
T	(superscript only) transpose of a vector or matrix
$\mathbf{x}$	column vector of variables, a point in $\mathbb{R}^n$ $\mathbf{x} = [x_1, x_2, \dots, x_n]^T$
$\in$	element in the set
$f(\mathbf{x}), f$	objective function
$\mathbf{x}^*$	local optimizer
$f(\mathbf{x}^*)$	optimum function value
$g_j(\mathbf{x}), g_j$	j^{th} inequality constraint function
$\mathbf{g}(\mathbf{x})$	vector of inequality constraint functions
$h_j(\mathbf{x}), h_j$	j^{th} equality constraint function
$\mathbf{h}(\mathbf{x})$	vector of equality constraint functions
C^1	set of continuous differentiable functions
C^2	set of continuous and twice continuous differentiable functions
$\min, \min\limits_{\mathbf{x}}$	minimize w.r.t. $\mathbf{x}$
$\mathbf{x}^0, \mathbf{x}^1, \dots$	vectors corresponding to points 0,1,...
$\{\mathbf{x} \mid \dots\}$	set of elements $\mathbf{x}$ such that ...
$\dfrac{\partial f}{\partial x_i}$	first partial derivative w.r.t. x_i
$\dfrac{\partial \mathbf{h}}{\partial x_i}$	$= [\dfrac{\partial h_1}{\partial x_i}, \dfrac{\partial h_2}{\partial x_i}, \dots, \dfrac{\partial h_r}{\partial x_i}]^T$
$\dfrac{\partial \mathbf{g}}{\partial x_i}$	$= [\dfrac{\partial g_1}{\partial x_i}, \dfrac{\partial g_2}{\partial x_i}, \dots, \dfrac{\partial g_m}{\partial x_i}]^T$
$\boldsymbol{\nabla}$	first derivative operator
$\boldsymbol{\nabla} f(\mathbf{x}) = \mathbf{g}(\mathbf{x})$	gradient vector $= \left[\dfrac{\partial f}{\partial x_1}(\mathbf{x}), \dfrac{\partial f}{\partial x_2}(\mathbf{x}), \dots, \dfrac{\partial f}{\partial x_n}(\mathbf{x})\right]^T$ (here $\mathbf{g}$ not to be confused with the inequality constraint function vector)

∇^2	second derivative operator (elements $\frac{\partial^2}{\partial x_i \partial x_j}$)
$\mathbf{H}(\mathbf{x}) = \nabla^2 f(\mathbf{x})$	Hessian matrix (second derivative matrix)
$\left.\frac{df(\mathbf{x})}{d\lambda}\right\|_{\mathbf{u}}$	directional derivative at $\mathbf{x}$ in the direction $\mathbf{u}$
$\subset, \subseteq$	subset of
$\lvert\cdot\rvert$	absolute value
$\lVert\cdot\rVert$	Euclidean norm of vector
$\cong$	approximately equal
$F(\,)$	line search function
$F[,]$	first order divided difference
$F[,,]$	second order divided difference
$(\mathbf{a}, \mathbf{b})$	scalar product of vector $\mathbf{a}$ and vector $\mathbf{b}$
$\mathbf{I}$	identity matrix
θ_j	j^{th} auxiliary variable
L	Lagrangian function
λ_j	j^{th} Lagrange multiplier
$\boldsymbol{\lambda}$	vector of Lagrange multipliers
$\exists$	exists
$\Rightarrow$	implies
$\{\cdots\}$	set
$V[\mathbf{x}]$	set of constraints violated at $\mathbf{x}$
ϕ	empty set
$\mathcal{L}$	augmented Lagrange function
$\langle a \rangle$	maximum of a and zero
$\frac{\partial \mathbf{h}}{\partial \mathbf{x}}$	$n \times r$ Jacobian matrix $= [\nabla h_1, \nabla h_2, \ldots, \nabla h_r]$
$\frac{\partial \mathbf{g}}{\partial \mathbf{x}}$	$n \times m$ Jacobian matrix $= [\nabla g_1, \nabla g_2, \ldots, \nabla g_m]$
s_i	slack variable
$\mathbf{s}$	vector of slack variables
D	determinant of matrix $\mathbf{A}$ of interest in $\mathbf{Ax} = \mathbf{b}$
D_j	determinant of matrix $\mathbf{A}$ with j^{th} column replaced by $\mathbf{b}$
$\lim_{i \to \infty}$	limit as i tends to infinity

Chapter 1

INTRODUCTION

1.1 What is mathematical optimization?

Formally, *Mathematical Optimization* is the process of

(i) the *formulation* and

(ii) the *solution* of a constrained optimization problem of the general mathematical form:

$$\underset{\text{w.r.t. } \mathbf{x}}{\text{minimize}}\, f(\mathbf{x}),\ \mathbf{x} = [x_1, x_2, \ldots, x_n]^T \in \mathbb{R}^n$$

subject to the constraints:

$$\begin{aligned} g_j(\mathbf{x}) &\leq 0, \quad j = 1,\ 2,\ \ldots,\ m \\ h_j(\mathbf{x}) &= 0, \quad j = 1,\ 2,\ \ldots,\ r \end{aligned} \tag{1.1}$$

where $f(\mathbf{x})$, $g_j(\mathbf{x})$ and $h_j(\mathbf{x})$ are scalar functions of the real *column vector* $\mathbf{x}$.

The continuous components x_i of $\mathbf{x} = [x_1, x_2, \ldots, x_n]^T$ are called the (*design*) *variables*, $f(\mathbf{x})$ is the *objective function*, $g_j(\mathbf{x})$ denotes the respective *inequality constraint functions* and $h_j(\mathbf{x})$ the *equality constraint functions.*

The optimum vector $\mathbf{x}$ that solves problem (1.1) is denoted by $\mathbf{x}^*$ with corresponding optimum function value $f(\mathbf{x}^*)$. If no constraints are specified, the problem is called an *unconstrained* minimization problem.

Mathematical Optimization is often also called *Nonlinear Programming, Mathematical Programming* or *Numerical Optimization.* In more general terms Mathematical Optimization may be described as the science of determining the *best* solutions to mathematically defined problems, which may be models of physical reality or of manufacturing and management systems. In the first case solutions are sought that often correspond to minimum energy configurations of general structures, from molecules to suspension bridges, and are therefore of interest to Science and Engineering. In the second case commercial and financial considerations of economic importance to Society and Industry come into play, and it is required to make decisions that will ensure, for example, maximum profit or minimum cost.

The history of the Mathematical Optimization, where functions of many variables are considered, is relatively short, spanning roughly only 55 years. At the end of the 1940s the very important simplex method for solving the special class of linear programming problems was developed. Since then numerous methods for solving the general optimization problem (1.1) have been developed, tested, and successfully applied to many important problems of scientific and economic interest. There is no doubt that the advent of the computer was essential for the development of these optimization methods. However, in spite of the proliferation of optimization methods, there is no universal method for solving all optimization problems. According to Nocedal and Wright (1999): "...there are numerous algorithms, each of which is tailored to a particular type of optimization problem. It is often the user's responsibility to choose an algorithm that is appropriate for the specific application. This choice is an important one; it may determine whether the problem is solved rapidly or slowly and, indeed, whether the solution is found at all." In a similar vein Vanderplaats (1998) states that "The author of each algorithm usually has numerical examples which demonstrate the efficiency and accuracy of the method, and the unsuspecting practitioner will often invest a great deal of time and effort in programming an algorithm, only to find that it will not in fact solve the particular problem being attempted. This often leads to disenchantment with these techniques that can be avoided if the user is knowledgeable in the basic concepts of numerical optimization." With these representative and authoritative opinions in mind, and also taking into account the present author's personal experiences in developing algorithms and applying them to design problems in mechanics, this text has been written to provide a brief but unified introduction to optimization concepts and methods. In addition, an overview of a set of novel algorithms, developed by the author and his students at the University of Pretoria over the past twenty years, is also given.

The emphasis of this book is almost exclusively on gradient-based methods. This is for two reasons. (i) The author believes that the introduction to the topic of mathematical optimization is best done via the classical gradient-based approach and (ii), contrary to the current popular trend of using non-gradient methods, such as genetic algorithms (GA's), simulated annealing, particle swarm optimization and other evolutionary methods, the author is of the opinion that these search methods are, in many cases, computationally too expensive to be viable. The argument that the presence of numerical noise and multiple minima disqualify the use of gradient-based methods, and that the only way out in such cases is the use of the above mentioned non-gradient search techniques, is not necessarily true. It is the experience of the author that, through the judicious use of gradient-based methods, problems with numerical noise and multiple minima may be solved, and at a fraction of the computational cost of search techniques such as genetic algorithms. In this context Chapter 4, dealing with the new gradient-based methods developed by the author, is especially important. The presentation of the material is not overly rigorous, but hopefully correct, and should provide the necessary information to allow scientists and engineers to select appropriate optimization algorithms and to apply them successfully to their respective fields of interest.

Many excellent and more comprehensive texts on practical optimization can be found in the literature. In particular the author wishes to acknowledge the works of Wismer and Chattergy (1978), Chvatel (1983), Fletcher (1987), Bazaraa et al. (1993), Arora (1989), Haftka and Gürdal (1992), Rao (1996), Vanderplaats (1998), Nocedal and Wright (1999) and Papalambros and Wilde (2000).

1.2 Objective and constraint functions

The values of the functions $f(\mathbf{x})$, $g_j(\mathbf{x})$ and $h_j(\mathbf{x})$ at any point $\mathbf{x} = [x_1, x_2, \dots, x_n]^T$, may in practice be obtained in different ways:

(i) from *analytically* known *formulae*, e.g. $f(\mathbf{x}) = x_1^2 + 2x_2^2 + \sin x_3$;

(ii) as the *outcome* of some complicated *computational process*, e.g. $g_1(\mathbf{x}) = a(\mathbf{x}) - a_{\text{max}}$, where $a(\mathbf{x})$ is the stress, computed by means of a finite element analysis, at some point in a structure, the design of which is specified by $\mathbf{x}$; or

(iii) from *measurements* taken of a *physical process*, e.g. $h_1(\mathbf{x}) = T(\mathbf{x}) - T_0$, where $T(\mathbf{x})$ is the temperature measured at some specified point in a reactor, and $\mathbf{x}$ is the vector of operational settings.

The first two ways of function evaluation are by far the most common. The optimization principles that apply in these cases, where computed function values are used, may be carried over directly to also be applicable to the case where the function values are obtained through physical measurements.

Much progress has been made with respect to methods for solving different classes of the general problem (1.1). Sometimes the solution may be obtained *analytically*, i.e. a closed-form solution in terms of a *formula* is obtained.

In general, especially for $n > 2$, solutions are usually obtained *numerically* by means of suitable *algorithms* (computational recipes).

Expertise in the *formulation* of appropriate optimization problems of the form (1.1), through which an optimum decision can be made, is gained from *experience*. This exercise also forms part of what is generally known as the *mathematical modelling* process. In brief, attempting to solve real-world problems via mathematical modelling requires the cyclic performance of the four steps depicted in Figure 1.1. The main steps are: 1) the observation and study of the real-world situation associated with a practical problem, 2) the abstraction of the problem by the construction of a mathematical model, that is described in terms

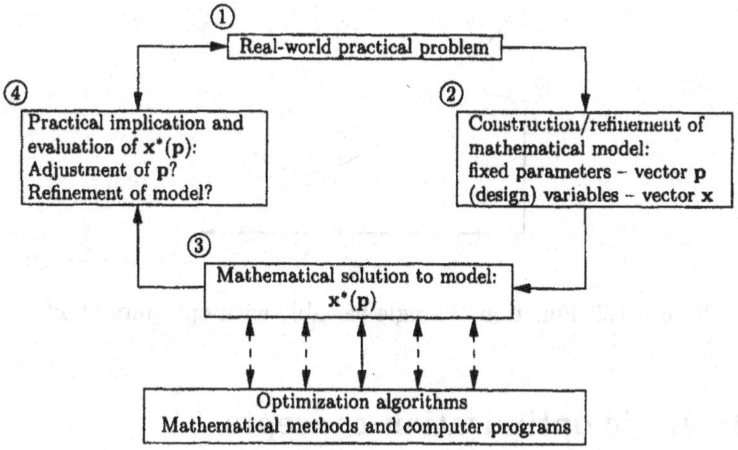

Figure 1.1: The mathematical modelling process

of preliminary fixed model parameters $\mathbf{p}$, and variables $\mathbf{x}$, the latter to be determined such that model performs in an acceptable manner, 3) the solution of a resulting purely mathematical problem, that requires an analytical or numerical parameter dependent solution $\mathbf{x}^*(\mathbf{p})$, and 4) the evaluation of the solution $\mathbf{x}^*(\mathbf{p})$ and its practical implications. After step 4) it may be necessary to adjust the parameters and refine the model, which will result in a new mathematical problem to be solved and evaluated. It may be required to perform the modelling cycle a number of times, before an acceptable solution is obtained. More often than not, the mathematical problem to be solved in 3) is a *mathematical optimization problem*, requiring a numerical solution. The *formulation* of an appropriate and consistent optimization problem (or model) is probably the most important, but unfortunately, also the *most neglected* part of Practical Mathematical Optimization.

This book gives a very brief introduction to the *formulation* of optimization problems, and deals with different *optimization algorithms* in greater depth. Since no algorithm is generally applicable to all classes of problems, the emphasis is on providing sufficient information to allow for the selection of appropriate algorithms or methods for different specific problems.

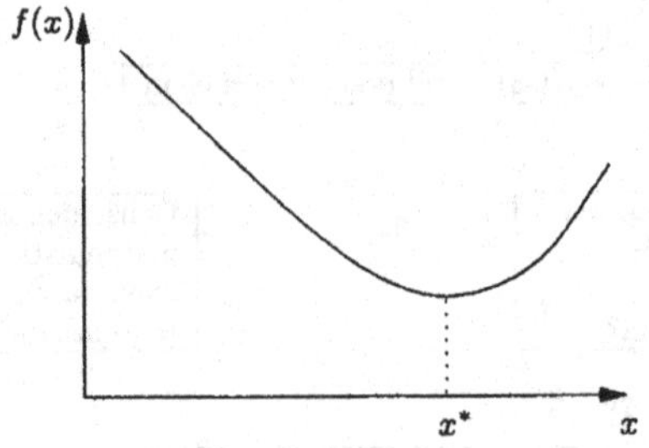

Figure 1.2: Function of single variable with optimum at x^*

1.3 Basic optimization concepts

1.3.1 Simplest class of problems: Unconstrained one-dimensional minimization

Consider the minimization of a smooth, i.e. continuous and twice continuously differentiable (C^2) function of a single real variable, i.e. the problem:

$$\underset{\text{w.r.t.}x}{\text{minimize}}\, f(x),\ x \in \mathbb{R},\ f \in C^2. \tag{1.2}$$

With reference to Figure 1.2, for a strong local minimum, it is required to determine a x^* such that $f(x^*) < f(x)$ for all x.

Clearly x^* occurs where the slope is zero, i.e. where

$$f'(x) = \frac{df(x)}{dx} = 0,$$

which corresponds to the first order necessary condition. In addition *non-negative curvature* is necessary at x^*, i.e. it is required that the second order condition

$$f''(x) = \frac{d^2f(x)}{dx^2} > 0$$

must hold at x^* for a strong local minimum.

A simple *special case* is where $f(x)$ has the simple *quadratic form*:

$$f(x) = ax^2 + bx + c. \tag{1.3}$$

Since the minimum occurs where $f'(x) = 0$, it follows that the closed-form solution is given by

$$x^* = -\frac{b}{2a}, \text{ provided } f''(x^*) = a > 0. \tag{1.4}$$

If $f(x)$ has a *more general form*, then a closed-form solution is in general not possible. In this case, the solution may be obtained numerically via the *Newton-Raphson algorithm:*

Given an approximation x^0, iteratively compute:

$$x^{i+1} = x^i - \frac{f'(x^i)}{f''(x^i)}; \; i = 0,\ 1,\ 2,\ \ldots \tag{1.5}$$

Hopefully $\lim_{i \to \infty} x^i = x^*$, i.e. the iterations converge, in which case a sufficiently accurate numerical solution is obtained after a finite number of iterations.

1.3.2 Contour representation of a function of two variables ($n = 2$)

Consider a function $f(\mathbf{x})$ of two variables, $\mathbf{x} = [x_1, x_2]^T$. The locus of all points satisfying $f(\mathbf{x}) = c =$ constant, forms a contour in the $x_1 - x_2$ plane. For each value of c there is a corresponding different contour.

Figure 1.3 depicts the contour representation for the example $f(\mathbf{x}) = x_1^2 + 2x_2^2$.

In three dimensions ($n = 3$), the contours are *surfaces of constant function* value. In more than three dimensions ($n > 3$) the contours are, of course, impossible to visualize. *Nevertheless, the contour representation in two-dimensional space will be used throughout the discussion of optimization techniques to help visualize the various optimization concepts.*

Other examples of 2-dimensional objective function contours are shown in Figures 1.4 to 1.6.

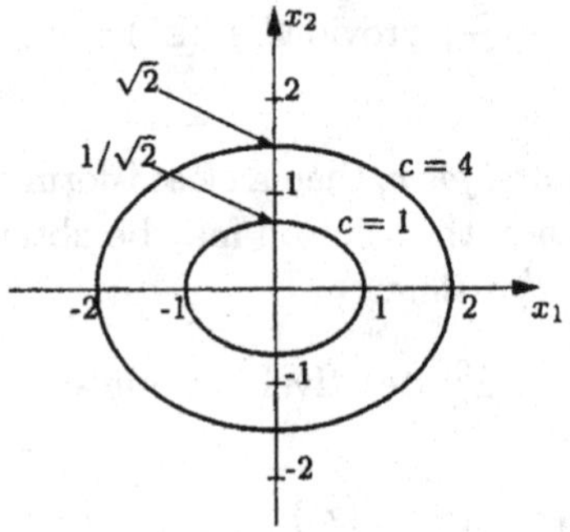

Figure 1.3: Contour representation of the function $f(\mathbf{x}) = x_1^2 + 2x_2^2$

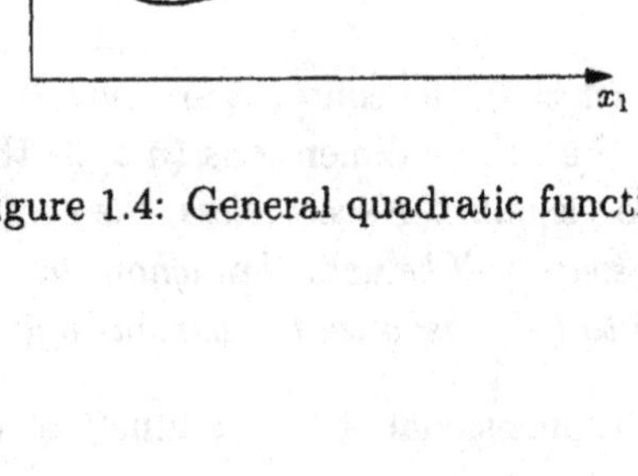

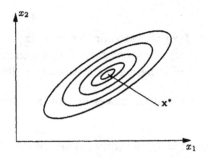

Figure 1.4: General quadratic function

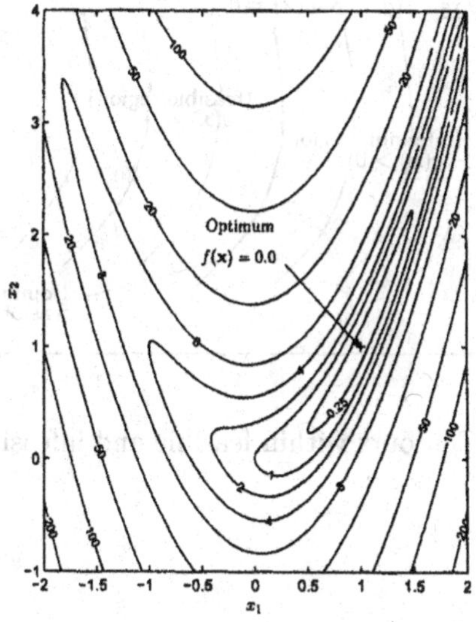

Figure 1.5: The 'banana' function $f(\mathbf{x}) = 10(x_2 - x_1^2)^2 + (1 - x_1)^2$

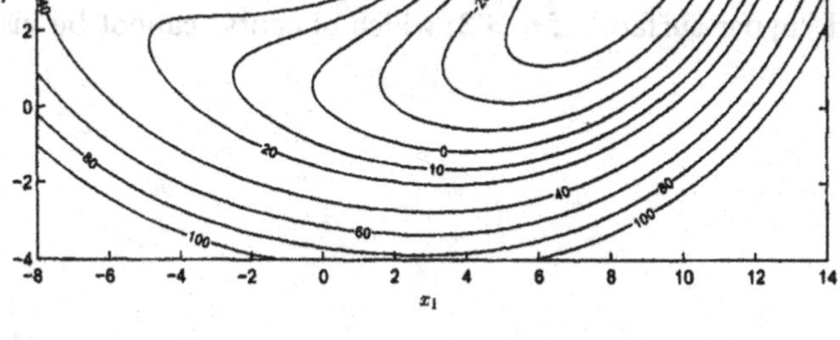

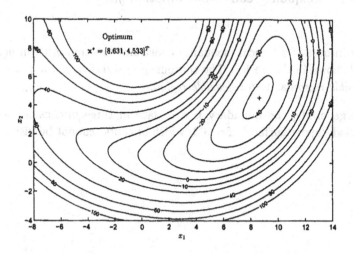

Figure 1.6: Potential energy function of a spring-force system (Vanderplaats, 1998)

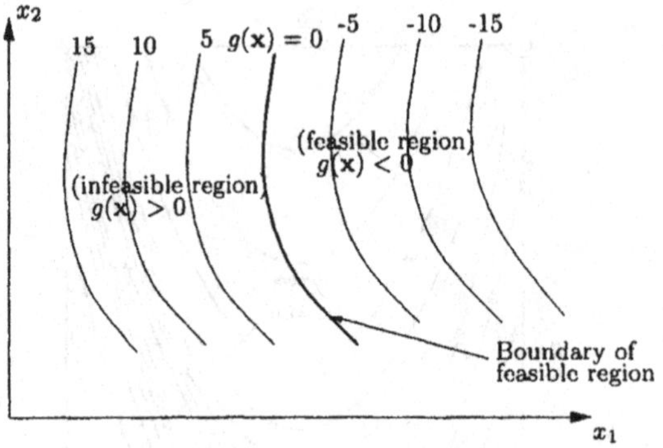

Figure 1.7: Contours within feasible and infeasible regions

1.3.3 Contour representation of constraint functions

1.3.3.1 Inequality constraint function $g(\mathbf{x})$

The contours of a typical inequality constraint function $g(\mathbf{x})$, in $g(\mathbf{x}) \leq 0$, are shown in Figure 1.7. The contour $g(\mathbf{x}) = 0$ divides the plane into a *feasible region* and an *infeasible region.*

More generally, the boundary is a surface in three dimensions and a so-called "hyper-surface" if $n > 3$, which of course cannot be visualised.

1.3.3.2 Equality constraint function $h(\mathbf{x})$

Here, as shown in Figure 1.8, only the line $h(\mathbf{x}) = 0$ is a feasible contour.

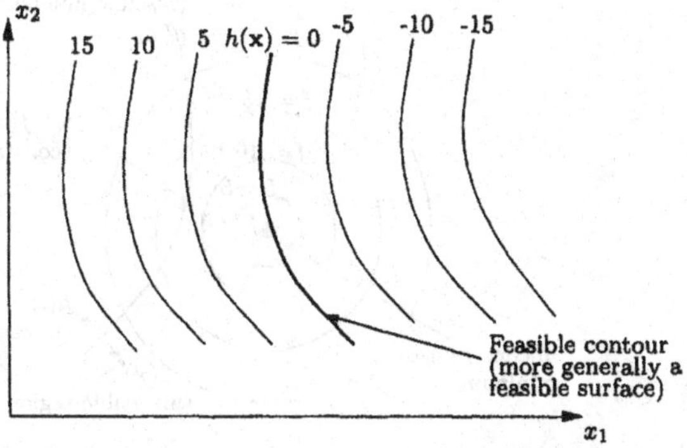

Figure 1.8: Feasible contour of equality constraint

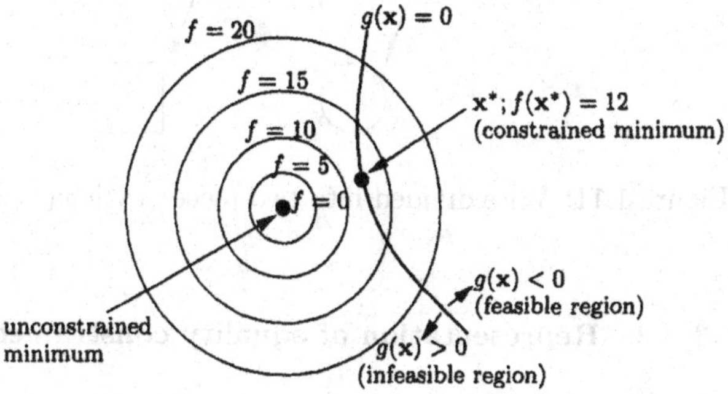

Figure 1.9: Contour representation of inequality constrained problem

1.3.4 Contour representations of constrained optimization problems

1.3.4.1 Representation of inequality constrained problem

Figure 1.9 graphically depicts the inequality constrained problem:

$$\min f(\mathbf{x})$$
$$\text{such that } g(\mathbf{x}) \leq 0.$$

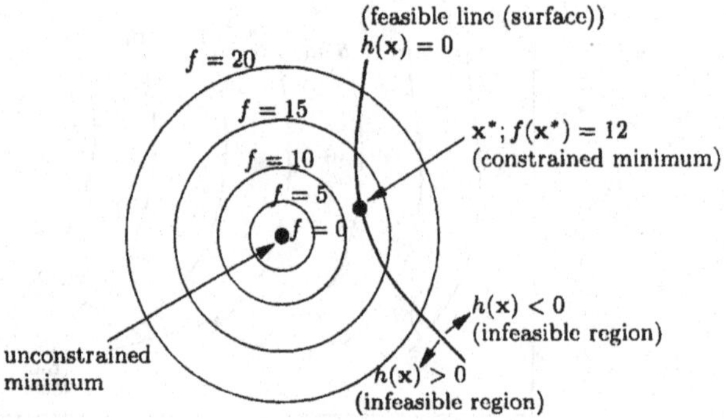

Figure 1.10: Contour representation of equality constrained problem

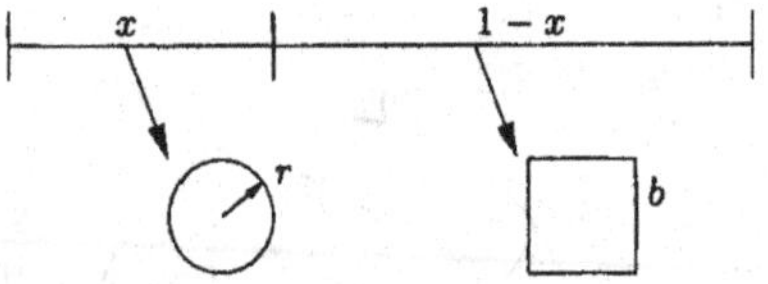

Figure 1.11: Wire divided into two pieces with $x_1 = x$ and $x_2 = 1 - x$

1.3.4.2 Representation of equality constrained problem

Figure 1.10 graphically depicts the equality constrained problem:

$$\begin{aligned} &\min f(\mathbf{x}) \\ \text{such that } h(\mathbf{x}) &= 0. \end{aligned}$$

1.3.5 Simple example illustrating the formulation and solution of an optimization problem

Problem: A length of wire 1 meter long is to be divided into two pieces, one in a circular shape and the other into a square as shown in Figure 1.11. What must the individual lengths be so that the total area is a minimum?

Formulation 1

Set length of first piece $= x$, then the area is given by $f(x) = \pi r^2 + b^2$. Since $r = \frac{x}{2\pi}$ and $b = \frac{1-x}{4}$ it follows that

$$f(x) = \pi\left(\frac{x^2}{4\pi^2}\right) + \frac{(1-x)^2}{16}.$$

The problem therefore reduces to an unconstrained minimization problem:

$$\text{minimize } f(x) = 0.1421x^2 - 0.125x + 0.0625.$$

Solution of Formulation 1

The function $f(x)$ is quadratic, therefore an analytical solution is given by the formula $x^* = -\frac{b}{2a}$ $(a > 0)$:

$$x^* = -\frac{-0.125}{2(0.1421)} = 0.4398 \text{ m},$$

and

$$1 - x^* = 0.5602 \text{ m with } f(x^*) = 0.0350 \text{ m}^2.$$

Formulation 2

Divide the wire into respective lengths x_1 and x_2 $(x_1 + x_2 = 1)$. The area is now given by

$$f(\mathbf{x}) = \pi r^2 + b^2 = \pi\left(\frac{x_1^2}{4\pi^2}\right) + \left(\frac{x_2}{4}\right)^2 = 0.0796x_1^2 + 0.0625x_2^2.$$

Here the problem reduces to an *equality constrained* problem:

$$\begin{aligned} \text{minimize } f(\mathbf{x}) &= 0.0796x_1^2 + 0.0625x_2^2 \\ \text{such that } h(\mathbf{x}) &= x_1 + x_2 - 1 = 0. \end{aligned}$$

Solution of Formulation 2

This constrained formulated problem is more difficult to solve. The closed-form analytical solution is not obvious and special constrained optimization techniques, such as the *method of Lagrange multipliers* to be discussed later, must be applied to solve the constrained problem analytically. The graphical solution is sketched in Figure 1.12.

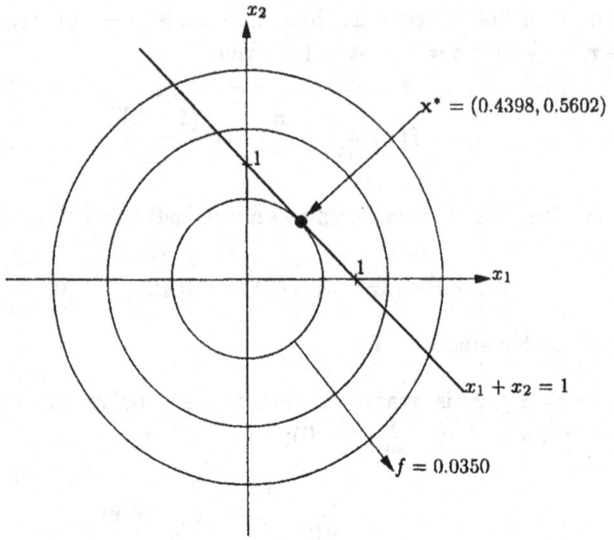

Figure 1.12: Graphical solution of Formulation 2

1.3.6 Maximization

The maximization problem: $\max_{\mathbf{x}} f(\mathbf{x})$ can be cast in the standard form (1.1) by observing that $\max_{\mathbf{x}} f(\mathbf{x}) = -\min_{\mathbf{x}}\{-f(\mathbf{x})\}$ as shown in Figure 1.13. Therefore in applying a minimization algorithm set $F(\mathbf{x}) = -f(\mathbf{x})$.

Also if the inequality constraints are given in the non-standard form: $g_j(\mathbf{x}) \geq 0$, then set $\tilde{g}_j(\mathbf{x}) = -g_j(\mathbf{x})$. In standard form the problem then becomes:

$$\text{minimize}\, F(\mathbf{x}) \text{ such that } \tilde{g}_j(\mathbf{x}) \leq 0.$$

Once the minimizer $\mathbf{x}^*$ is obtained, the maximum value of the original maximization problem is given by $-F(\mathbf{x}^*)$.

1.3.7 The special case of Linear Programming

A very important special class of the general optimization problem arises when both the objective function and all the constraints are linear func-

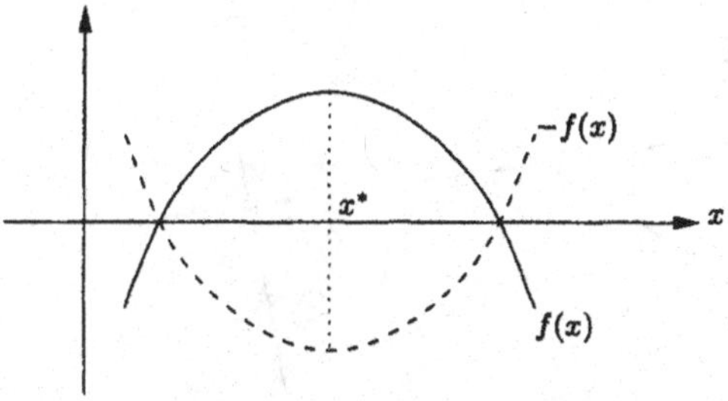

Figure 1.13: Maximization problem transformed to minimization problem

tions of $\mathbf{x}$. This is called a *Linear Programming* problem and is usually stated in the following form:

$$\begin{aligned} &\min_{\mathbf{x}} f(\mathbf{x}) = \mathbf{c}^T\mathbf{x} \\ &\text{such that} \qquad (1.6) \\ &\mathbf{Ax} \leq \mathbf{b};\ \mathbf{x} \geq \mathbf{0} \end{aligned}$$

where $\mathbf{c}$ is a real n-vector and $\mathbf{b}$ is a real m-vector, and $\mathbf{A}$ is a $m \times n$ real matrix. A linear programming problem in two variables is graphically depicted in Figure 1.14.

Special methods have been developed for solving linear programming problems. Of these the most famous are the simplex method proposed by Dantzig in 1947 (Dantzig, 1963) and the interior-point method (Karmarkar, 1984). A short introduction to the simplex method, according to Chvatel (1983), is given in Appendix A.

1.3.8 Scaling of design variables

In formulating mathematical optimization problems great care must be taken to ensure that the scale of the variables are more or less of the same order. If not, the formulated problem may be relatively insensitive to the variations in one or more of the variables, and any optimization algorithm will struggle to converge to the true solution, because of extreme distortion of the objective function contours as result of the poor scaling. In particular it may lead to difficulties when selecting step lengths and

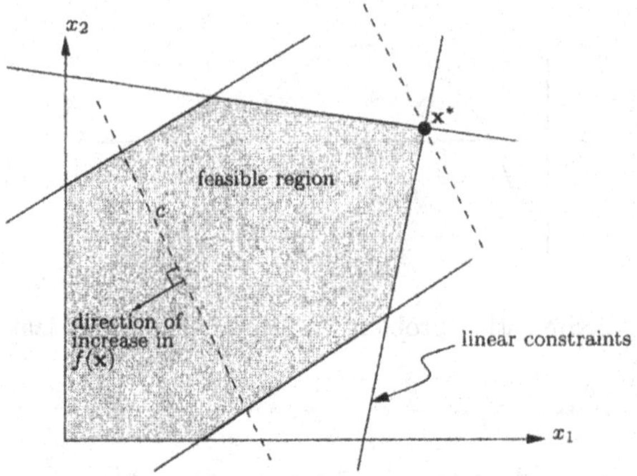

Figure 1.14: Graphical representation of a two-dimensional linear programming problem

calculating numerical gradients. Scaling difficulties often occur where the variables are of different dimension and expressed in different units. Hence it is good practice, if the variable ranges are very large, to scale the variables so that all the variables will be dimensionless and vary between 0 and 1 approximately. For scaling the variables, it is necessary to establish an approximate range for each of the variables. For this, take some estimates (based on judgement and experience) for the lower and upper limits. The values of the bounds are not critical. Another related matter is the scaling or normalization of constraint functions. This becomes necessary whenever the values of the constraint functions differ by large magnitudes.

1.4 Further mathematical prerequisites

1.4.1 Convexity

A line through the points $\mathbf{x}^1$ and $\mathbf{x}^2$ in $\mathbb{R}^n$ is the set

$$L = \{\mathbf{x} | \mathbf{x} = \mathbf{x}^1 + \lambda(\mathbf{x}^2 - \mathbf{x}^1), \text{ for all } \lambda \in \mathbb{R}\}. \tag{1.7}$$

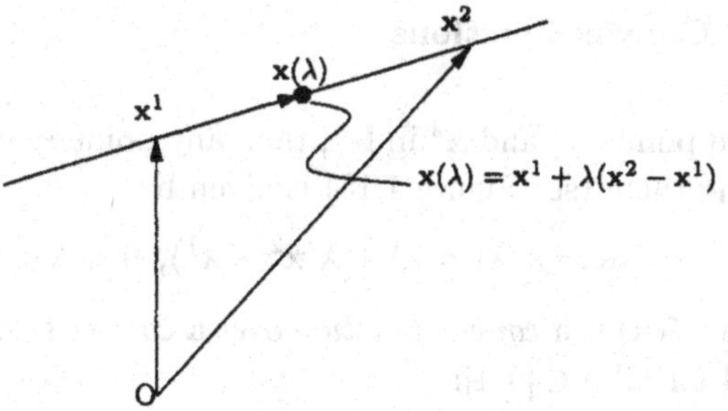

Figure 1.15: Representation of a point on the straight line through $\mathbf{x}^1$ and $\mathbf{x}^2$

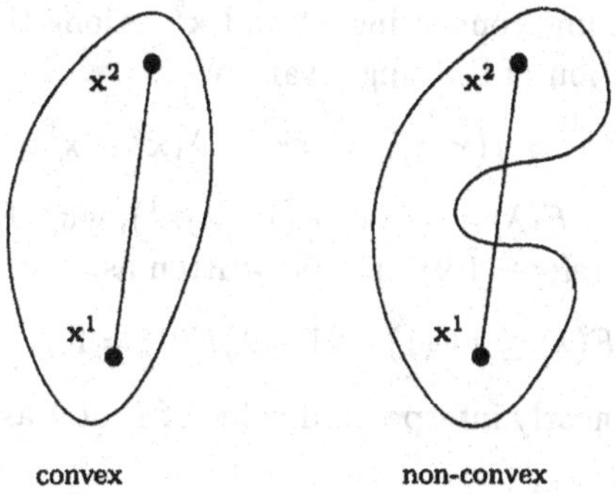

Figure 1.16: Examples of a convex and a non-convex set

Equivalently for any point $\mathbf{x}$ on the line there exists a λ such that $\mathbf{x}$ may be specified by $\mathbf{x} = \mathbf{x}(\lambda) = \lambda\mathbf{x}^2 + (1 - \lambda)\mathbf{x}^1$ as shown in Figure 1.15.

1.4.1.1 Convex sets

A set X is convex if for all $\mathbf{x}^1,\ \mathbf{x}^2 \in X$ it follows that

$$\mathbf{x} = \lambda\mathbf{x}^2 + (1 - \lambda)\mathbf{x}^1 \in X \text{ for all } 0 \leq \lambda \leq 1.$$

If this condition does not hold the set is non-convex (see Figure 1.16).

1.4.1.2 Convex functions

Given two points $\mathbf{x}^1$ and $\mathbf{x}^2$ in $\mathbb{R}^n$, then any point $\mathbf{x}$ on the straight line connecting them (see Figure 1.15) is given by

$$\mathbf{x} = \mathbf{x}(\lambda) = \mathbf{x}^1 + \lambda(\mathbf{x}^2 - \mathbf{x}^1),\ 0 < \lambda < 1. \tag{1.8}$$

A function $f(\mathbf{x})$ is a *convex function* over a *convex set* X if for all $\mathbf{x}^1$, $\mathbf{x}^2$ in X and for all $\lambda \in [0, 1]$:

$$f(\lambda\mathbf{x}^2 + (1-\lambda)\mathbf{x}^1) \leq \lambda f(\mathbf{x}^2) + (1-\lambda)f(\mathbf{x}^1). \tag{1.9}$$

The function is strictly convex if $<$ applies. Concave functions are similarly defined.

Consider again the line connecting $\mathbf{x}^1$ and $\mathbf{x}^2$. Along this line, the function $f(\mathbf{x})$ is a function of the single variable λ:

$$F(\lambda) = f(\mathbf{x}(\lambda)) = f(\mathbf{x}^1 + \lambda(\mathbf{x}^2 - \mathbf{x}^1)). \tag{1.10}$$

This is equivalent to $F(\lambda) = f(\lambda\mathbf{x}^2 + (1-\lambda)\mathbf{x}^1)$, with $F(0) = f(\mathbf{x}^1)$ and $F(1) = f(\mathbf{x}^2)$. Therefore (1.9) may be written as

$$F(\lambda) \leq \lambda F(1) + (1-\lambda)F(0) = F_{int}$$

where F_{int} is the linearly interpolated value of F at λ as shown in Figure 1.17.

Graphically $f(\mathbf{x})$ is *convex* over the convex set X if $F(\lambda)$ has the convex form shown in Figure 1.17 for any two points $\mathbf{x}^1$ and $\mathbf{x}^2$ in X.

1.4.2 Gradient vector of $f(\mathbf{x})$

For a function $f(\mathbf{x}) \in C^2$ there exists, at any point $\mathbf{x}$ a vector of first order partial derivatives, or gradient vector:

$$\nabla f(\mathbf{x}) = \begin{bmatrix} \frac{\partial f}{\partial x_1}(\mathbf{x}) \\ \frac{\partial f}{\partial x_2}(\mathbf{x}) \\ \vdots \\ \frac{\partial f}{\partial x_n}(\mathbf{x}) \end{bmatrix} = \mathbf{g}(\mathbf{x}). \tag{1.11}$$

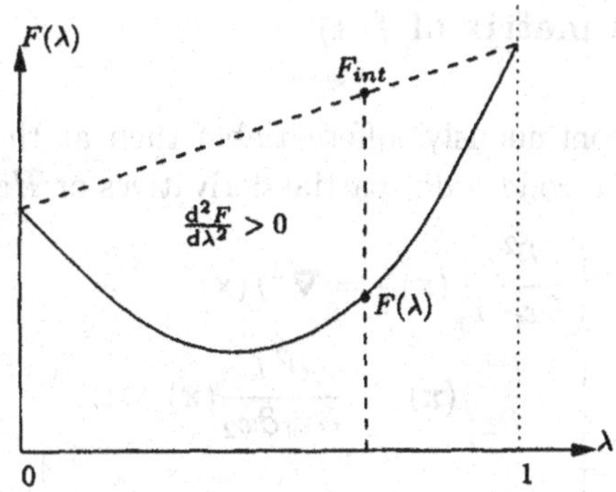

Figure 1.17: Convex form of $F(\lambda)$

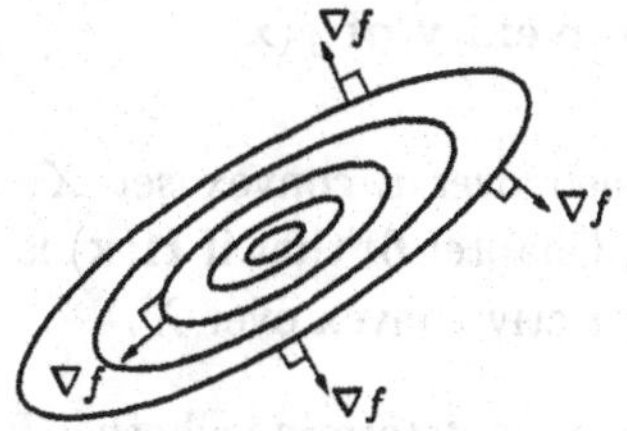

Figure 1.18: Directions of the gradient vector

It can easily be shown that if the function $f(\mathbf{x})$ is smooth, then at the point $\mathbf{x}$ the gradient vector $\boldsymbol{\nabla} f(\mathbf{x})$ (also often denoted by $\mathbf{g}(\mathbf{x})$) is always perpendicular to the contours (or surfaces of constant function value) and is in the *direction of maximum increase* of $f(\mathbf{x})$, as depicted in Figure 1.18.

1.4.3 Hessian matrix of $f(\mathbf{x})$

If $f(\mathbf{x})$ is twice continuously differentiable then at the point $\mathbf{x}$ there exists a matrix of second order partial derivatives or *Hessian matrix*:

$$\mathbf{H}(\mathbf{x}) = \left\{\frac{\partial^2 f}{\partial x_i \partial x_j}(\mathbf{x})\right\} = \nabla^2 f(\mathbf{x}) \tag{1.12}$$

$$= \begin{bmatrix} \frac{\partial^2 f}{\partial x_1^2}(\mathbf{x}) & \frac{\partial^2 f}{\partial x_1 \partial x_2}(\mathbf{x}) & \dots & \\ \frac{\partial^2 f}{\partial x_2 \partial x_1}(\mathbf{x}) & & & \\ \vdots & & & \\ \frac{\partial^2 f}{\partial x_n \partial x_1}(\mathbf{x}) & & \dots & \frac{\partial^2 f}{\partial x_n^2}(\mathbf{x}) \end{bmatrix}.$$

Clearly $\mathbf{H}(\mathbf{x})$ is a $n \times n$ symmetrical matrix.

1.4.3.1 Test for convexity of $f(\mathbf{x})$

If $f(\mathbf{x}) \in C^2$ is defined over a convex set X, then it can be shown (see Theorem 6.1.3 in Chapter 6) that if $\mathbf{H}(\mathbf{x})$ is positive-definite for all $\mathbf{x} \in X$, then $f(\mathbf{x})$ is strictly convex over X.

To test for convexity, i.e. to determine whether $\mathbf{H}(\mathbf{x})$ is positive-definite or not, apply Sylvester's Theorem or any other suitable numerical method (Fletcher, 1987).

1.4.4 The quadratic function in $\mathbb{R}^n$

The quadratic function in n variables may be written as

$$f(\mathbf{x}) = \tfrac{1}{2}\mathbf{x}^T\mathbf{A}\mathbf{x} + \mathbf{b}^T\mathbf{x} + c \tag{1.13}$$

where $c \in \mathbb{R}$, $\mathbf{b}$ is a real n-vector and $\mathbf{A}$ is a $n \times n$ real matrix that can be chosen in a non-unique manner. It is usually chosen symmetrical in which case it follows that

$$\nabla f(\mathbf{x}) = \mathbf{A}\mathbf{x} + \mathbf{b}; \ \mathbf{H}(\mathbf{x}) = \mathbf{A}. \tag{1.14}$$

The function $f(\mathbf{x})$ is called positive-definite if $\mathbf{A}$ is positive-definite since, by the test in 1.4.3.1, a function $f(\mathbf{x})$ is convex if $\mathbf{H}(\mathbf{x})$ is positive-definite.

1.4.5 The directional derivative of $f(\mathbf{x})$ in the direction $\mathbf{u}$

It is usually assumed that $\|\mathbf{u}\| = 1$. Consider the differential:

$$df = \frac{\partial f}{\partial x_1}dx_1 + \cdots + \frac{\partial f}{\partial x_n}dx_n = \nabla^T f(\mathbf{x})d\mathbf{x}. \tag{1.15}$$

A point $\mathbf{x}$ on the line through $\mathbf{x}'$ in the direction $\mathbf{u}$ is given by $\mathbf{x} = \mathbf{x}(\lambda) = \mathbf{x}' + \lambda\mathbf{u}$, and for a small change $d\lambda$ in λ, $d\mathbf{x} = \mathbf{u}d\lambda$. Along this line $F(\lambda) = f(\mathbf{x}' + \lambda\mathbf{u})$ and the differential at any point $\mathbf{x}$ on the given line in the direction $\mathbf{u}$ is therefore given by $dF = df = \nabla^T f(\mathbf{x})\mathbf{u}d\lambda$. It follows that the *directional derivative* at $\mathbf{x}$ in the *direction* $\mathbf{u}$ is

$$\frac{dF(\lambda)}{d\lambda} = \left.\frac{df(\mathbf{x})}{d\lambda}\right|_{\mathbf{u}} = \nabla^T f(\mathbf{x})\mathbf{u}. \tag{1.16}$$

1.5 Unconstrained minimization

In considering the unconstrained problem: $\min_{\mathbf{x}} f(\mathbf{x}),\ \mathbf{x} \in X \subseteq \mathbb{R}^n$, the following questions arise:

(i) what are the conditions for a minimum to exist,

(ii) is the minimum unique,

(iii) are there any relative minima?

Figure 1.19 (after Farkas and Jarmai, 1997) depicts different types of minima that may arise for functions of a single variable, and for functions of two variables in the presence of equality constraints. Intuitively, with reference to Figure 1.19, one feels that a general function may have a single unique global minimum, or it may have more than one local minimum. The function may indeed have no local minimum at all, and in two dimensions the possibility of saddle points also comes to mind.

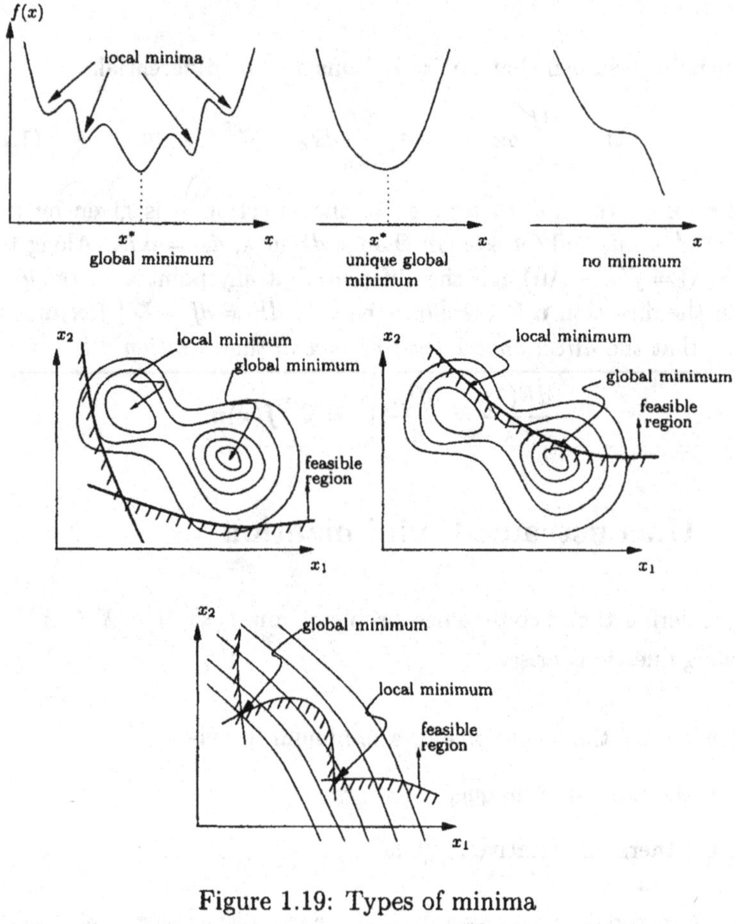

Figure 1.19: Types of minima

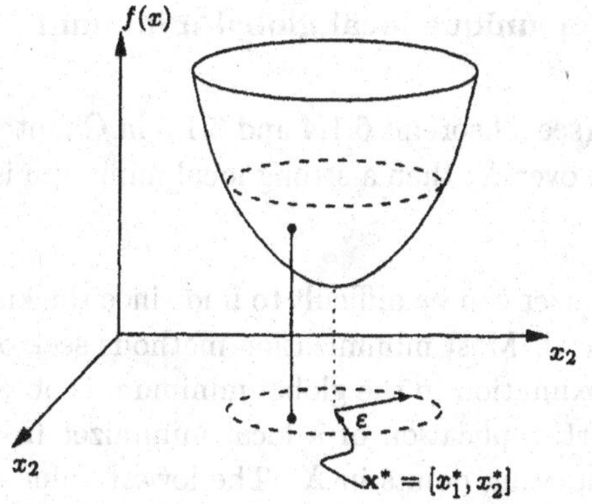

Figure 1.20: Graphical representation of the definition of a local minimum

Thus, in order to answer the above questions regarding the nature of any given function more analytically, it is necessary to give more precise meanings to the above mentioned notions.

1.5.1 Global and local minima; saddle points

1.5.1.1 Global minimum

$\mathbf{x}^*$ is a global minimum over the set X if $f(\mathbf{x}) \geq f(\mathbf{x}^*)$ for all $\mathbf{x} \in X \subset \mathbb{R}^n$.

1.5.1.2 Strong local minimum

$\mathbf{x}^*$ is a strong local minimum if there exists an $\varepsilon > 0$ such that

$$f(\mathbf{x}) > f(\mathbf{x}^*) \text{ for all } \{\mathbf{x} \big| \|\mathbf{x} - \mathbf{x}^*\| < \varepsilon\}$$

where $\|\cdot\|$ denotes the Euclidean norm. This definition is sketched in Figure 1.20.

1.5.1.3 Test for unique local global minimum

It can be shown (see Theorems 6.1.4 and 6.1.5 in Chapter 6) that if $f(\mathbf{x})$ is strictly convex over X, then a strong local minimum is also the global minimum.

The global minimizer can be difficult to find since the knowledge of $f(\mathbf{x})$ is usually only local. Most minimization methods seek only a local minimum. An approximation to the global minimum is obtained in practice by the multi-start application of a local minimizer from randomly selected different starting points in X. The lowest value obtained after a sufficient number of trials is then taken as a good approximation to the global solution (see Snyman and Fatti, 1987; Groenwold and Snyman, 2002). If, however, it is known that the function is strictly convex over X, then only one trial is sufficient since only one local minimum, the global minimum, exists.

1.5.1.4 Saddle points

$f(\mathbf{x})$ has a saddle point at $\overline{\mathbf{x}} = \begin{bmatrix} \mathbf{x}^0 \\ \mathbf{y}^0 \end{bmatrix}$ if there exists an $\varepsilon > 0$ such that for all $\mathbf{x}$, $\|\mathbf{x} - \mathbf{x}^0\| < \varepsilon$ and all $\mathbf{y}$, $\|\mathbf{y} - \mathbf{y}^0\| < \varepsilon$: $f(\mathbf{x}, \mathbf{y}^0) \leq f(\mathbf{x}^0, \mathbf{y}^0) \leq f(\mathbf{x}^0, \mathbf{y})$.

A contour representation of a saddle point in two dimensions is given in Figure 1.21.

1.5.2 Local characterization of the behaviour of a multi-variable function

It is assumed here that $f(\mathbf{x})$ is a smooth function, i.e., that it is a twice continuously differentiable function ($f(\mathbf{x}) \in C^2$). Consider again the line $\mathbf{x} = \mathbf{x}(\lambda) = \mathbf{x}' + \lambda\mathbf{u}$ through the point $\mathbf{x}'$ in the direction $\mathbf{u}$.

Along this line a single variable function $F(\lambda)$ may be defined:

$$F(\lambda) = f(\mathbf{x}(\lambda)) = f(\mathbf{x}' + \lambda\mathbf{u}).$$

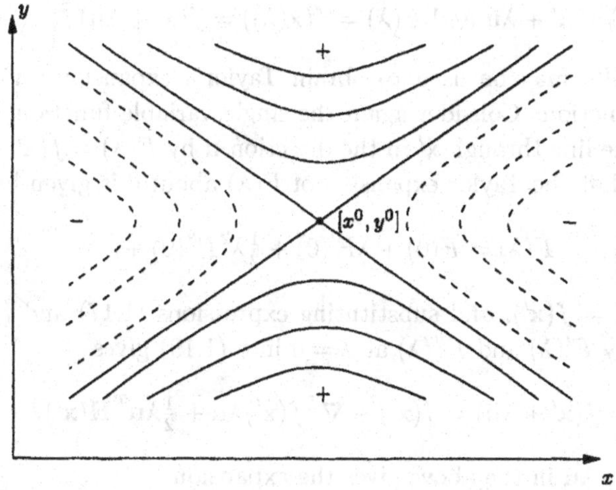

Figure 1.21: Contour representation of saddle point

It follows from (1.16) that

$$\frac{dF(\lambda)}{d\lambda} = \left.\frac{df(\mathbf{x}(\lambda))}{d\lambda}\right|_{\mathbf{u}} = \boldsymbol{\nabla}^T f(\mathbf{x}(\lambda))\mathbf{u} = g(\mathbf{x}(\lambda)) = G(\lambda)$$

which is also a single variable function of λ along the line $\mathbf{x} = \mathbf{x}(\lambda) = \mathbf{x}' + \lambda\mathbf{u}$.

Thus similarly it follows that

$$\begin{aligned}\frac{d^2F(\lambda)}{d\lambda^2} = \frac{dG(\lambda)}{d\lambda} = \left.\frac{dg(\mathbf{x}(\lambda))}{d\lambda}\right|_{\mathbf{u}} &= \boldsymbol{\nabla}^T g(\mathbf{x}(\lambda))\mathbf{u} \\ &= \boldsymbol{\nabla}^T\left(\boldsymbol{\nabla}^T f(\mathbf{x}(\lambda))\mathbf{u}\right)\mathbf{u} \\ &= \mathbf{u}^T\mathbf{H}(\mathbf{x}(\lambda))\mathbf{u}.\end{aligned}$$

Summarising: the first and second order derivatives of $F(\lambda)$ with respect to λ at any point $\mathbf{x} = \mathbf{x}(\lambda)$ on any line (any $\mathbf{u}$) through $\mathbf{x}'$ is given by

$$\frac{dF(\lambda)}{d\lambda} = \boldsymbol{\nabla}^T f(\mathbf{x}(\lambda))\mathbf{u}, \tag{1.17}$$

$$\frac{d^2F(\lambda)}{d\lambda^2} = \mathbf{u}^T\mathbf{H}(\mathbf{x}(\lambda))\mathbf{u} \tag{1.18}$$

where $\mathbf{x}(\lambda) = \mathbf{x}' + \lambda\mathbf{u}$ and $F(\lambda) = f(\mathbf{x}(\lambda)) = f(\mathbf{x}' + \lambda\mathbf{u})$.

These results may be used to obtain Taylor's expansion for a multi-variable function. Consider again the single variable function $F(\lambda)$ defined on the line through $\mathbf{x}'$ in the direction $\mathbf{u}$ by $F(\lambda) = f(\mathbf{x}' + \lambda\mathbf{u})$. It is known that the Taylor expansion of $F(\lambda)$ about 0 is given by

$$F(\lambda) = F(0) + \lambda F'(0) + \tfrac{1}{2}\lambda^2 F''(0) + \dots \tag{1.19}$$

With $F(0) = f(\mathbf{x}')$, and substituting expressions (1.17) and (1.18) for respectively $F'(\lambda)$ and $F''(\lambda)$ at $\lambda = 0$ into (1.19) gives

$$F(\lambda) = f(\mathbf{x}' + \lambda\mathbf{u}) = f(\mathbf{x}') + \boldsymbol{\nabla}^T f(\mathbf{x}')\lambda\mathbf{u} + \tfrac{1}{2}\lambda\mathbf{u}^T\mathbf{H}(\mathbf{x}')\lambda\mathbf{u} + \dots$$

Setting $\boldsymbol{\delta} = \lambda\mathbf{u}$ in the above gives the expansion:

$$f(\mathbf{x}' + \boldsymbol{\delta}) = f(\mathbf{x}') + \boldsymbol{\nabla}^T f(\mathbf{x}')\boldsymbol{\delta} + \tfrac{1}{2}\boldsymbol{\delta}^T\mathbf{H}(\mathbf{x}')\boldsymbol{\delta} + \dots \tag{1.20}$$

Since the above applies for any line (any $\mathbf{u}$) through $\mathbf{x}'$, it represents the general Taylor expansion for a multi-variable function about $\mathbf{x}'$. If $f(\mathbf{x})$ is fully continuously differentiable in the neighbourhood of $\mathbf{x}'$ it can be shown that the truncated second order Taylor expansion for a multi-variable function is given by

$$f(\mathbf{x}' + \boldsymbol{\delta}) = f(\mathbf{x}') + \boldsymbol{\nabla}^T f(\mathbf{x}')\boldsymbol{\delta} + \tfrac{1}{2}\boldsymbol{\delta}^T\mathbf{H}(\mathbf{x}' + \theta\boldsymbol{\delta})\boldsymbol{\delta} \tag{1.21}$$

for some $\theta \in [0, 1]$. This expression is important in the analysis of the behaviour of a multi-variable function at any given point $\mathbf{x}'$.

1.5.3 Necessary and sufficient conditions for a strong local minimum at $\mathbf{x}^*$

In particular, consider $\mathbf{x}' = \mathbf{x}^*$ a strong local minimizer. Then for any line (any $\mathbf{u}$) through $\mathbf{x}'$ the behaviour of $F(\lambda)$ in a neighbourhood of $\mathbf{x}^*$ is as shown in Figure 1.22, with minimum at at $\lambda = 0$.

Clearly, a *necessary first order condition* that must apply at $\mathbf{x}^*$ (corresponding to $\lambda = 0$) is that

$$\frac{dF(0)}{d\lambda} = \boldsymbol{\nabla}^T f(\mathbf{x}^*)\mathbf{u} = 0, \quad \text{for all } \mathbf{u} \neq \mathbf{0}. \tag{1.22}$$

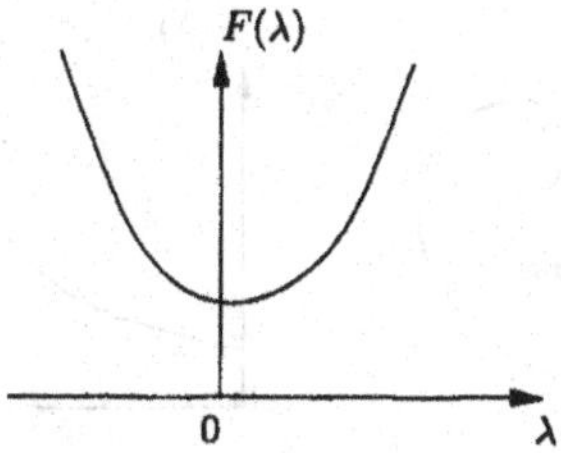

Figure 1.22: Behaviour of $F(\lambda)$ near $\lambda = 0$

It can easily be shown that this condition also implies that necessarily $\nabla f(\mathbf{x}^*) = \mathbf{0}$.

A *necessary second order condition* that must apply at $\mathbf{x}^*$ is that

$$\frac{d^2F(0)}{d\lambda^2} = \mathbf{u}^T\mathbf{H}(\mathbf{x}^*)\mathbf{u} > 0, \quad \text{for all } \mathbf{u} \neq \mathbf{0}. \tag{1.23}$$

Conditions (1.22) and (1.23) taken together are also *sufficient conditions* (i.e. those that imply) for $\mathbf{x}^*$ to be a strong local minimum if $f(\mathbf{x})$ is continuously differentiable in the vicinity of $\mathbf{x}^*$. This can easily be shown by substituting these conditions in the Taylor expansion (1.21).

Thus in summary, the *necessary and sufficient conditions* for $\mathbf{x}^*$ to be a strong local minimum are:

$$\begin{array}{l} \nabla f(\mathbf{x}^*) = \mathbf{0} \\ \mathbf{H}(\mathbf{x}^*) \text{ positive-definite.} \end{array} \tag{1.24}$$

In the argument above it has implicitly been assumed that $\mathbf{x}^*$ is an unconstrained minimum *interior* to X. If $\mathbf{x}^*$ lies on the *boundary* of X (see Figure 1.23) then

$$\frac{dF(0)}{d\lambda} \geq 0, \text{ i.e. } \nabla^T f(\mathbf{x}^*)\mathbf{u} \geq 0 \tag{1.25}$$

for all *allowable* directions $\mathbf{u}$, i.e. for directions such that $\mathbf{x}^* + \lambda\mathbf{u} \in X$ for arbitrary small $\lambda > 0$.

Conditions (1.24) for an unconstrained strong local minimum play a very important role in the construction of practical algorithms for unconstrained optimization.

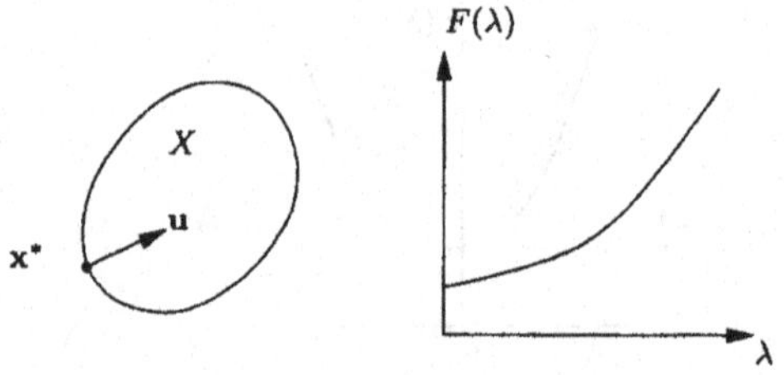

Figure 1.23: Behaviour of $F(\lambda)$ for all allowable directions of u

1.5.3.1 Application to the quadratic function

Consider the quadratic function:

$$f(\mathbf{x}) = \tfrac{1}{2}\mathbf{x}^T\mathbf{A}\mathbf{x} + \mathbf{b}^T\mathbf{x} + c.$$

In this case the first order necessary condition for a minimum implies that

$$\nabla f(\mathbf{x}) = \mathbf{A}\mathbf{x} + \mathbf{b} = \mathbf{0}.$$

Therefore a candidate solution point is

$$\mathbf{x}^* = -\mathbf{A}^{-1}\mathbf{b}. \tag{1.26}$$

If the second order necessary condition also applies, i.e. if $\mathbf{A}$ is positive-definite, then $\mathbf{x}^*$ is a unique minimizer.

1.5.4 General indirect method for computing $\mathbf{x}^*$

The general indirect method for determining $\mathbf{x}^*$ is to solve the system of equations $\nabla f(\mathbf{x}) = \mathbf{0}$ (corresponding to the first order necessary condition in(1.24)) by some numerical method, to yield all stationary points. An obvious method for doing this is Newton's method. Since in general the system will be non-linear, multiple stationary points are possible. These stationary points must then be further analysed in order to determine whether or not they are local minima.

1.5.4.1 Solution by Newton's method

Assume $\mathbf{x}^*$ is a local minimum and $\mathbf{x}^i$ an approximate solution, with associated unknown error $\boldsymbol{\delta}$ such that $\mathbf{x}^* = \mathbf{x}^i + \boldsymbol{\delta}$. Then by applying Taylor's theorem and the first order necessary condition for a minimum at $\mathbf{x}^*$ it follows that

$$\mathbf{0} = \nabla f(\mathbf{x}^*) = \nabla f(\mathbf{x}^i + \boldsymbol{\delta}) = \nabla f(\mathbf{x}^i) + \mathbf{H}(\mathbf{x}^i)\boldsymbol{\delta} + \mathrm{O}\|\boldsymbol{\delta}\|^2.$$

If $\mathbf{x}^i$ is a good approximation then $\boldsymbol{\delta} \doteq \boldsymbol{\Delta}$, the solution of the linear system $\mathbf{H}(\mathbf{x}^i)\boldsymbol{\Delta} + \nabla f(\mathbf{x}^i) = \mathbf{0}$, obtained by ignoring the second order term in $\boldsymbol{\delta}$ above. A better approximation is therefore expected to be $\mathbf{x}^{i+1} = \mathbf{x}^i + \boldsymbol{\Delta}$ which leads to the Newton iterative scheme: Given an initial approximation $\mathbf{x}^0$, compute

$$\mathbf{x}^{i+1} = \mathbf{x}^i - \mathbf{H}^{-1}(\mathbf{x}^i)\nabla f(\mathbf{x}^i) \tag{1.27}$$

for $i = 0,\ 1,\ 2,\ \ldots$ Hopefully $\lim_{i\to\infty} \mathbf{x}^i = \mathbf{x}^*$.

1.5.4.2 Example of Newton's method applied to a quadratic problem

Consider the unconstrained problem:

$$\text{minimize } f(\mathbf{x}) = \tfrac{1}{2}\mathbf{x}^T\mathbf{A}\mathbf{x} + \mathbf{b}^T\mathbf{x} + c.$$

In this case the first iteration in (1.27) yields

$$\mathbf{x}^1 = \mathbf{x}^0 - \mathbf{A}^{-1}(\mathbf{A}\mathbf{x}^0 + \mathbf{b}) = \mathbf{x}^0 - \mathbf{x}^0 - \mathbf{A}^{-1}\mathbf{b} = -\mathbf{A}^{-1}\mathbf{b}$$

i.e. $\mathbf{x}^1 = \mathbf{x}^* = -\mathbf{A}^{-1}\mathbf{b}$ in a single step (see (1.26)). This is to be expected since in this case no approximation is involved and thus $\boldsymbol{\Delta} = \boldsymbol{\delta}$.

1.5.4.3 Difficulties with Newton's method

Unfortunately, in spite of the attractive features of the Newton method, such as being quadratically convergent near the solution, the basic Newton method as described above does not always perform satisfactorily. The main difficulties are:

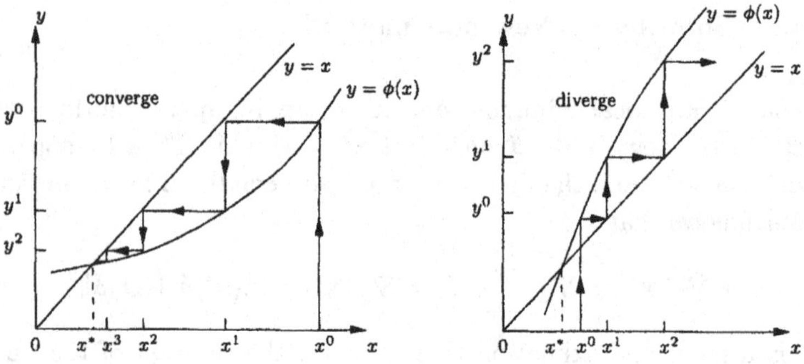

Figure 1.24: Graphical representation of Newton's iterative scheme for a single variable

(i) the method is not always convergent, even if $\mathbf{x}^0$ is close to $\mathbf{x}^*$, and

(ii) the method requires the computation of the Hessian matrix at each iteration.

The first of these difficulties may be illustrated by considering Newton's method applied to the one-dimensional problem: solve $f'(x) = 0$. In this case the iterative scheme is

$$x^{i+1} = x^i - \frac{f'(x^i)}{f''(x^i)} = \phi(x^i), \text{ for } i = 0,\ 1,\ 2,\ \ldots \tag{1.28}$$

and the solution corresponds to the fixed point x^* where $x^* = \phi(x^*)$. Unfortunately in some cases, unless x^0 is chosen to be exactly equal to x^*, convergence will not necessarily occur. In fact, convergence is dependent on the nature of the fixed point function $\phi(x)$ in the vicinity of x^*, as shown for two different ϕ functions in Figure 1.24. With reference to the graphs Newton's method is: $y^i = \phi(x^i)$, $x^{i+1} = y^i$ for $i = 0,\ 1,\ 2,\ \ldots$. Clearly in the one case where $|\phi'(x)| < 1$ convergence occurs, but in the other case where $|\phi'(x)| > 1$ the scheme diverges.

In more dimensions the situation may be even more complicated. In addition, for a large number of variables, difficulty (ii) mentioned above becomes serious in that the computation of the Hessian matrix represents a major task. If the Hessian is not available in analytical form, use can be made of automatic differentiation techniques to compute it, or it can be estimated by means of finite differences. It should also be noted that in computing the Newton step in (1.27) a $n \times n$ linear system must be solved. This represents further computational effort. Therefore in practice the simple basic Newton method is not recommended. To avoid the convergence difficulty use is made of a modified Newton method, in which a more direct search procedure is employed in the direction of the Newton step, so as to ensure descent to the minimum $\mathbf{x}^*$. The difficulty associated with the computation of the Hessian is addressed in practice through the systematic update, from iteration to iteration, of an approximation to the Hessian matrix. These improvements to the basic Newton method are dealt with in greater detail in the next chapter.

1.6 Exercises

1.6.1 Sketch the graphical solution to the following problem:

$$\min_{\mathbf{x}} f(\mathbf{x}) = (x_1 - 2)^2 + (x_2 - 2)^2$$
$$\text{such that } x_1 + 2x_2 = 4;\ x_1 \geq 0;\ x_2 \geq 0.$$

In particular indicate the feasible region: $F = \{(x_1, x_2) | x_1 + 2x_2 = 4; x_1 \geq 0; x_2 \geq 0\}$ and the solution point $\mathbf{x}^*$.

1.6.2 Show that x^2 is a convex function.

1.6.3 Show that the sum of convex functions is also convex.

1.6.4 Determine the gradient vector and Hessian of the Rosenbrock function:

$$f(\mathbf{x}) = 100(x_2 - x_1^2)^2 + (1 - x_1)^2.$$

1.6.5 Write the quadratic function $f(\mathbf{x}) = x_1^2 + 2x_1x_2 + 3x_2^2$ in the standard matrix-vector notation. Is $f(\mathbf{x})$ positive-definite?

1.6.6 Write each of the following objective functions in standard form:

$$f(\mathbf{x}) = \tfrac{1}{2}\mathbf{x}^T\mathbf{A}\mathbf{x} + \mathbf{b}^T\mathbf{x} + c.$$

(i) $f(\mathbf{x}) = x_1^2 + 2x_1x_2 + 4x_1x_3 + 3x_2^2 + 2x_2x_3 + 5x_3^2 + 4x_1 - 2x_2 + 3x_3$.

Chapter 2

LINE SEARCH DESCENT METHODS FOR UNCONSTRAINED MINIMIZATION

2.1 General line search descent algorithm for unconstrained minimization

Over the last 40 years many powerful *direct search algorithms* have been developed for the unconstrained minimization of general functions. These algorithms require an initial estimate to the optimum point, denoted by $\mathbf{x}^0$. With this estimate as starting point, the algorithm generates a sequence of estimates $\mathbf{x}^0, \mathbf{x}^1, \mathbf{x}^2, \ldots$, by successively searching *directly* from each point in a direction of *descent* to determine the next point. The process is terminated if either no further progress is made, or if a point $\mathbf{x}^k$ is reached (for smooth functions) at which the first necessary condition in (1.24), i.e. $\nabla f(\mathbf{x}) = \mathbf{0}$ is sufficiently accurately satisfied, in which case $\mathbf{x}^* \cong \mathbf{x}^k$. It is usually, although not always, required that the function value at the new iterate $\mathbf{x}^{i+1}$ be lower than that at $\mathbf{x}^i$.

An important sub-class of direct search methods, specifically suitable for smooth functions, are the so-called *line search* descent methods. Basic to these methods is the selection of a descent direction $\mathbf{u}^{i+1}$ at each iterate $\mathbf{x}^i$ that ensures descent at $\mathbf{x}^i$ in the direction $\mathbf{u}^{i+1}$, i.e. it is required that the directional derivative in the direction $\mathbf{u}^{i+1}$ be negative:

$$\left.\frac{df(\mathbf{x}^i)}{d\lambda}\right|_{\mathbf{u}^{i+1}} = \boldsymbol{\nabla}^T f(\mathbf{x}^i)\mathbf{u}^{i+1} < 0. \tag{2.1}$$

The general structure of such descent methods is given below.

2.1.1 General structure of a line search descent method

1. Given starting point $\mathbf{x}^0$ and positive tolerances ε_1, ε_2 and ε_3, set $i = 1$.

2. Select a descent direction $\mathbf{u}^i$ (see descent condition (2.1).

3. Perform a *one-dimensional line search* in direction $\mathbf{u}^i$: i.e.

$$\min_{\lambda} F(\lambda) = \min_{\lambda} f(\mathbf{x}^{i-1} + \lambda\mathbf{u}^i)$$

   to give minimizer λ_i.

4. Set $\mathbf{x}^i = \mathbf{x}^{i-1} + \lambda_i\mathbf{u}^i$.

5. Test for convergence:

   if $\|\mathbf{x}^i - \mathbf{x}^{i-1}\| < \varepsilon_1$, or $\|\boldsymbol{\nabla} f(\mathbf{x}^i)\| < \varepsilon_2$, or $|f(\mathbf{x}^i) - f(\mathbf{x}^{i-1})| < \varepsilon_3$, then STOP and $\mathbf{x}^* \cong \mathbf{x}^i$,

   else go to Step 6.

6. Set $i = i + 1$ and go to Step 2.

In testing for termination in step 5, a combination of the stated termination criteria may be used, i.e. instead of *or*, *and* may be specified. The structure of the above descent algorithm is depicted in Figure 2.1.

Different descent methods, within the above sub-class, differ according to the way in which the descent directions $\mathbf{u}^i$ are chosen. Another important consideration is the method by means of which the one-dimensional line search is performed.

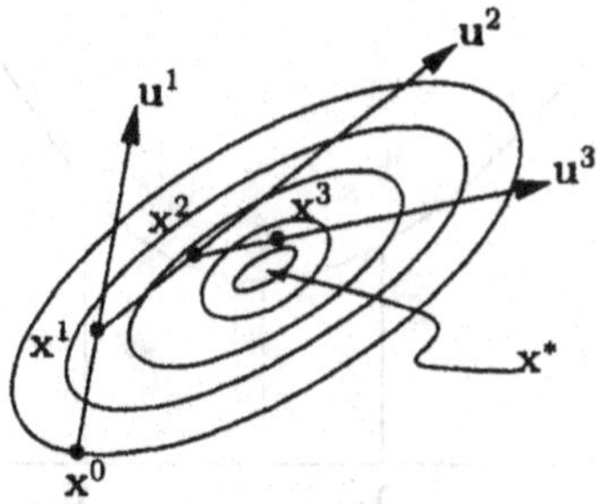

Figure 2.1: Sequence of line search descent directions and steps

2.2 One-dimensional line search

Clearly, in implementing descent algorithms of the above type, the one-dimensional minimization problem:

$$\min_{\lambda} F(\lambda), \ \lambda \in \mathbb{R} \tag{2.2}$$

is an important sub-problem. Here the minimizer is denoted by λ^*, i.e.

$$F(\lambda^*) = \min_{\lambda} F(\lambda).$$

Many one-dimensional minimization techniques have been proposed and developed over the years. These methods differ according to whether they are to be applied to smooth functions or poorly conditioned functions. For smooth functions *interpolation methods*, such as the quadratic interpolation method of Powell (1964) and the cubic interpolation algorithm of Davidon (1959), are the most efficient and accurate methods. For poorly conditioned functions, *bracketing methods*, such as the Fibonacci search method (Kiefer, 1957), which is optimal with respect to the number of function evaluations required for a prescribed accuracy, and the golden section method (Walsh, 1975), which is near optimal but much simpler and easier to implement, are preferred. Here *Powell's quadratic interpolation* method and the *golden section* method, are respectively presented as representative of the two different approaches that may be adopted to one-dimensional minimization.

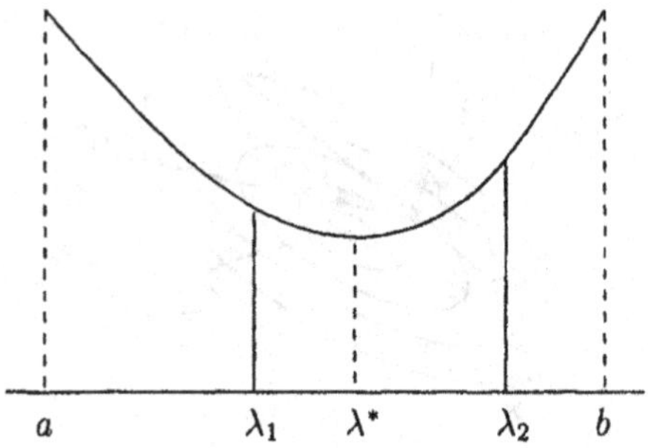

Figure 2.2: Unimodal function $F(\lambda)$ over interval $[a, b]$

2.2.1 Golden section method

It is assumed that $F(\lambda)$ is *unimodal* over the interval $[a, b]$, i.e. that it has a minimum λ^* within the interval and that $F(\lambda)$ is strictly descending for $\lambda < \lambda^*$ and strictly ascending for $\lambda > \lambda^*$, as shown in Figure 2.2.

Note that if $F(\lambda)$ is unimodal over $[a, b]$ with λ^* in $[a, b]$, then to determine a sub-unimodal interval, at least *two* evaluations of $F(\lambda)$ in $[a, b]$ must be made as indicated in Figure 2.2.

If $F(\lambda_2) > F(\lambda_1) \Rightarrow$ new unimodal interval $= [a, \lambda_2]$, and set $b = \lambda_2$ and select new λ_2; otherwise new unimodal interval $= [\lambda_1, b]$ and set $a = \lambda_1$ and select new λ_1.

Thus, the unimodal interval may successively be reduced by inspecting values of $F(\lambda_1)$ and $F(\lambda_2)$ at interior points λ_1 and λ_2.

The question arises: How can λ_1 and λ_2 be chosen in the most economic manner, i.e. such that a least number of function evaluations are required for a prescribed accuracy (i.e. for a specified uncertainty interval)? The most economic method is the Fibonacci search method. It is however a complicated method. A near optimum and more straightforward method is the golden section method. This method is a limiting form of the Fibonacci search method. Use is made of the golden ratio r when selecting the values for λ_1 and λ_2 within the unimodal interval. The value of r corresponds to the positive root of the quadratic equation: $r^2 + r - 1 = 0$, thus $r = \frac{\sqrt{5}-1}{2} = 0.618034$.

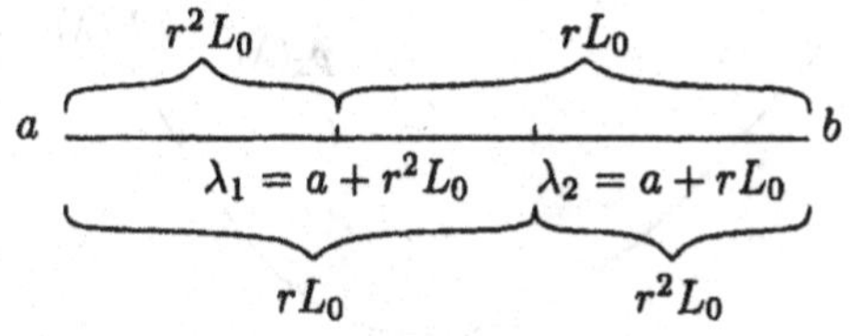

Figure 2.3: Selection of interior points λ_1 and λ_2 for golden section search

The details of the selection procedure are as follows. Given initial unimodal interval $[a, b]$ of length L_0, then choose interior points λ_1 and λ_2 as shown in Figure 2.3.

Then, if $F(\lambda_1) > F(\lambda_2) \Rightarrow$ new $[a, b] = [\lambda_1, b]$ with new interval length $L_1 = rL_0$, and

if $F(\lambda_2) > F(\lambda_1) \Rightarrow$ new $[a, b] = [a, \lambda_2]$ also with $L_1 = rL_0$.

The detailed formal algorithm is stated below.

2.2.1.1 Basic golden section algorithm

Given interval $[a, b]$ and prescribed accuracy ε; then set $i = 0$; $L_0 = b - a$, and perform the following steps:

1. Set $\lambda_1 = a + r^2L_0$; $\lambda_2 = a + rL_0$.
2. Compute $F(\lambda_1)$ and $F(\lambda_2)$; set $i = i + 1$.
3. *If* $F(\lambda_1) > F(\lambda_2)$ then

   set $a = \lambda_1$; $\lambda_1 = \lambda_2$; $L_i = (b - a)$; and $\lambda_2 = a + rL_i$,

   else

   set $b = \lambda_2$; $\lambda_2 = \lambda_1$; $L_i = (b - a)$; and $\lambda_1 = a + r^2L_i$.
4. *If* $L_i < \varepsilon$ then

   set $\lambda^* = \dfrac{b + a}{2}$; compute $F(\lambda^*)$ and STOP,

   else go to Step 2.

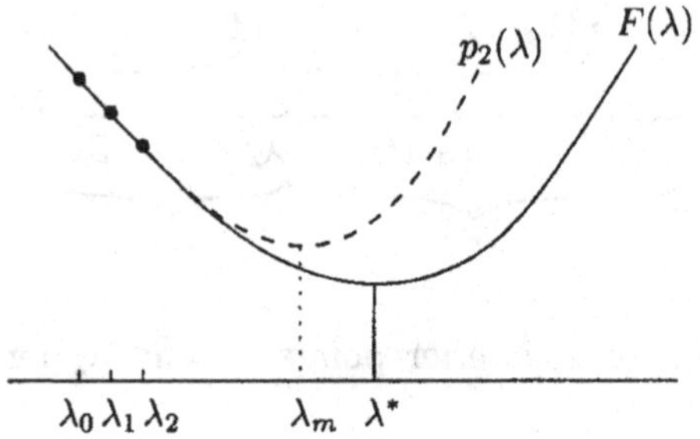

Figure 2.4: Approximate minimum λ_m via quadratic interpolation

2.2.2 Powell's quadratic interpolation algorithm

In Powell's method successive quadratic interpolation curves are fitted to function data giving a sequence of approximations to the minimum point λ^*.

With reference to Figure 2.4, the basic idea is the following. Given three data points $\{(\lambda_i, F(\lambda_i)),\ i = 1, 2, 3\}$, then the interpolating quadratic polynomial through these points $p_2(\lambda)$ is given by

$$p_2(\lambda) = F(\lambda_0) + F[\lambda_0, \lambda_1](\lambda - \lambda_0) + F[\lambda_0, \lambda_1, \lambda_2](\lambda - \lambda_0)(\lambda - \lambda_1) \quad (2.3)$$

where $F[\ ,\ ]$ and $F[\ ,\ ,\ ]$ respectively denote the first order and second order divided differences.

The turning point of $p_2(\lambda)$ occurs where the slope is zero, i.e. where

$$\frac{dp_2}{d\lambda} = F[\lambda_0, \lambda_1] + 2\lambda F[\lambda_0, \lambda_1, \lambda_2] - F[\lambda_0, \lambda_1, \lambda_2](\lambda_0 + \lambda_1) = 0$$

which gives the turning point λ_m as

$$\lambda_m = \frac{F[\lambda_0, \lambda_1, \lambda_2](\lambda_0 + \lambda_1) - F[\lambda_0, \lambda_1]}{2F[\lambda_0, \lambda_1, \lambda_2]} \cong \lambda^* \quad (2.4)$$

with the further condition that for a minimum the second derivative must be non-negative, i.e. $F[\lambda_0, \lambda_1, \lambda_2] > 0$.

The detailed formal algorithm is as follows.

2.2.2.1 Powell's interpolation algorithm

Given starting point λ_0, stepsize h, tolerance ε and maximum stepsize H; perform following steps:

1. Compute $F(\lambda_0)$ and $F(\lambda_0 + h)$.
2. *If* $F(\lambda_0) < F(\lambda_0 + h)$ evaluate $F(\lambda_0 - h)$,

   else evaluate $F(\lambda_0 + 2h)$. (The three initial values of λ so chosen constitute the initial set $(\lambda_0, \lambda_1, \lambda_2)$ with corresponding function values $F(\lambda_i)$, $i = 0, 1, 2$.)
3. Compute turning point λ_m by formula (2.4) and test for minimum or maximum.
4. *If* λ_m a minimum point *and* $|\lambda_m - \lambda_n| > H$, where λ_n is the nearest point to λ_m, then discard the point furthest from λ_m and take a step of size H from the point with lowest value in direction of descent, and go to Step 3;

   if λ_m a maximum point, then discard point nearest λ_m and take a step of size H from the point with lowest value in the direction of descent and go to Step 3;

   else continue.
5. *If* $|\lambda_m - \lambda_n| < \varepsilon$ then $F(\lambda^*) \cong \min[F(\lambda_m), F(\lambda_n)]$ and STOP,

   else continue.
6. Discard point with highest F value and replace it by λ_m; go to Step 3

Note: It is always safer to compute the next turning point by interpolation rather than by extrapolation. Therefore in Step 6: if the maximum value of F corresponds to a point which lies alone on one side of λ_m, then rather discard the point with highest value on the other side of λ_m.

2.2.3 Exercises

Apply the golden section method and Powell's method to the problems below. Compare their respective performances with regard to the number of function evaluation required to attain the prescribed accuracies.

(i) minimize $F(\lambda) = \lambda^2 + 2e^{-\lambda}$ over $[0, 2]$ with $\varepsilon = 0.01$.

(ii) maximize $F(\lambda) = \lambda \cos \lambda$ over $[0, \pi/2]$ with $\varepsilon = 0.001$.

(iii) minimize $F(\lambda) = 4(\lambda-7)/(\lambda^2+\lambda-2)$ over $[-1.9; 0.9]$ by performing only 10 function evaluations.

(iv) minimize $F(\lambda) = \lambda^4 - 20\lambda^3 + 0.1\lambda$ over $[0; 20]$ with $\varepsilon = 10^{-5}$.

2.3 First order line search descent methods

Line search descent methods (see Section 2.1.1), that use the gradient vector $\boldsymbol{\nabla} f(\mathbf{x})$ to determine the search direction for each iteration, are called *first order methods* because they employ *first order* partial derivatives of $f(\mathbf{x})$ to compute the search direction at the current iterate. The simplest and most famous of these methods is the *method of steepest descent*, first proposed by Cauchy in 1847.

2.3.1 The method of steepest descent

In this method the direction of steepest descent is used as the search direction in the line search descent algorithm given in section 2.1.1. The expression for the direction of steepest descent is derived below.

2.3.1.1 The direction of steepest descent

At $\mathbf{x}'$ we seek the unit vector $\mathbf{u}$ such that for $F(\lambda) = f(\mathbf{x}' + \lambda\mathbf{u})$, the directional derivative

$$\left.\frac{df(\mathbf{x}')}{d\lambda}\right|_{\mathbf{u}} = \frac{dF(0)}{d\lambda} = \boldsymbol{\nabla}^T f(\mathbf{x}')\mathbf{u}$$

assumes a minimum value with respect to all possible choices for the unit vector $\mathbf{u}$ at $\mathbf{x}'$ (see Figure 2.5).

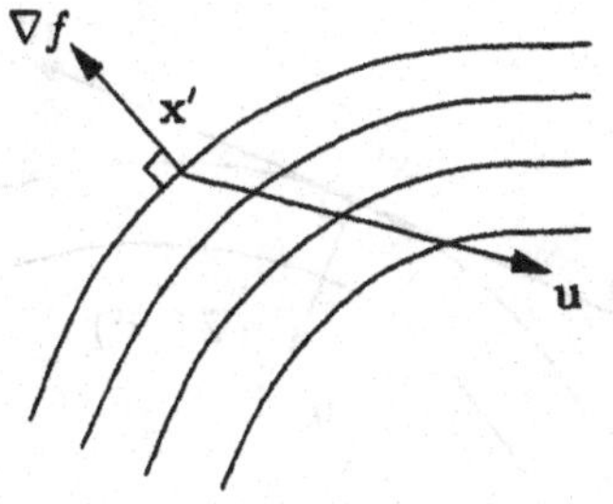

Figure 2.5: Search direction $\mathbf{u}$ relative to gradient vector at $\mathbf{x}'$

By Schwartz's inequality:

$$\nabla^T f(\mathbf{x}')\mathbf{u} \geq -\|\nabla f(\mathbf{x}')\|\|\mathbf{u}\| = -\|\nabla f(\mathbf{x}')\| = \text{ least value.}$$

Clearly for the particular choice $\mathbf{u} = \dfrac{-\nabla f(\mathbf{x}')}{\|\nabla f(\mathbf{x}')\|}$ the directional derivative at $\mathbf{x}'$ is given by

$$\frac{dF(0)}{d\lambda} = -\nabla^T f(\mathbf{x}')\frac{\nabla f(\mathbf{x}')}{\|\nabla f(\mathbf{x}')\|} = -\|\nabla f(\mathbf{x}')\| = \text{ least value.}$$

Thus this particular choice for the unit vector corresponds to the direction of steepest descent.

The search direction

$$\mathbf{u} = \frac{-\nabla f(\mathbf{x})}{\|\nabla f(\mathbf{x})\|} \tag{2.5}$$

is called the ***normalized steepest descent direction*** at $\mathbf{x}$.

2.3.1.2 Steepest descent algorithm

Given $\mathbf{x}^0$, do for iteration $i = 1, 2, \ldots$ until convergence:

1. set $\mathbf{u}^i = \dfrac{-\nabla f(\mathbf{x}^{i-1})}{\|\nabla f(\mathbf{x}^{i-1})\|}$

2. set $\mathbf{x}^i = \mathbf{x}^{i-1} + \lambda_i \mathbf{u}^i$ where λ_i is such that

$$F(\lambda_i) = f(\mathbf{x}^{i-1} + \lambda_i \mathbf{u}^i) = \min_{\lambda} f(\mathbf{x}^{i-1} + \lambda \mathbf{u}^i) \text{ (line search).}$$

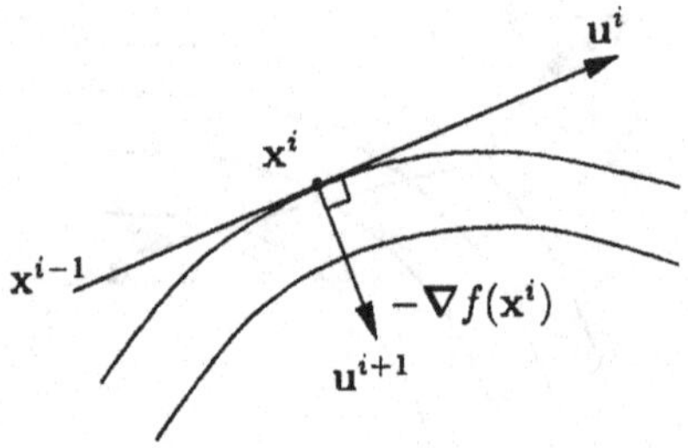

Figure 2.6: Orthogonality of successive steepest descent search directions

2.3.1.3 Characteristic property

Successive steepest descent search directions can be shown to be *orthogonal.* Consider the line search through $\mathbf{x}^{i-1}$ in the direction $\mathbf{u}^i$ to give $\mathbf{x}^i$. The condition for a minimum at λ_i, i.e. for optimal descent, is

$$\left.\frac{df(\mathbf{x}^{i-1}+\lambda_i\mathbf{u}^i)}{d\lambda}\right|_{\mathbf{u}^i} = \left.\frac{dF(\lambda_i)}{d\lambda}\right|_{\mathbf{u}^i} = \nabla^T f(\mathbf{x}^i)\mathbf{u}^i = 0$$

and with $\mathbf{u}^{i+1} = -\dfrac{\nabla f(\mathbf{x}^i)}{\|\nabla f(\mathbf{x}^i)\|}$ it follows that $\mathbf{u}^{i+1^T}\mathbf{u}^i = 0$ as shown in Figure 2.6.

2.3.1.4 Convergence criteria

In practice the algorithm is terminated if some convergence criterion is satisfied. Usually termination is enforced at iteration i if one, or a combination, of the following criteria is met:

(i) $\|\mathbf{x}^i - \mathbf{x}^{i-1}\| < \varepsilon_1$

(ii) $\|\nabla f(\mathbf{x}^i)\| < \varepsilon_2$

(iii) $|f(\mathbf{x}^i) - f(\mathbf{x}^{i-1})| < \varepsilon_3$.

where ε_1, ε_2 and ε_3 are prescribed small positive tolerances.

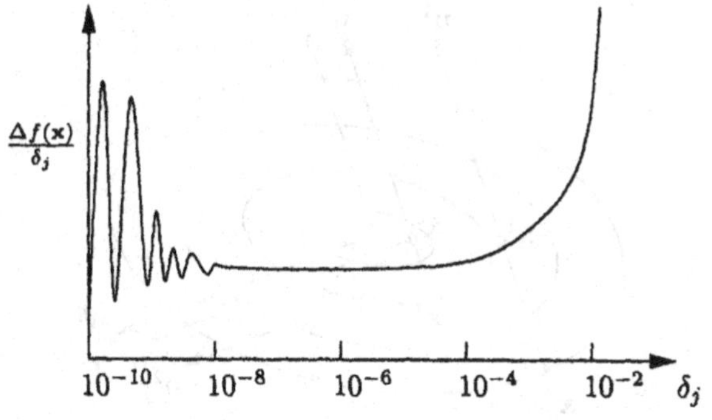

Figure 2.7: Sensitivity of finite difference approximation to δ_j

2.3.1.5 Gradients by finite differences

Often the components of the gradient vector is not analytically available in which case they may be approximated by forward finite differences:

$$\frac{\partial f(\mathbf{x})}{\partial x_j} \cong \frac{\Delta f(\mathbf{x})}{\delta_j} = \frac{f(\mathbf{x}+\boldsymbol{\delta}_j) - f(\mathbf{x})}{\delta_j} \tag{2.6}$$

where $\boldsymbol{\delta}_j = [0, 0, \ldots \delta_j, 0, \ldots, 0]^T$, $\delta_j > 0$ in the j-th position.

Often $\delta_j \equiv \delta$ for all $j = 1, 2, \ldots, n$. A typically choice is $\delta = 10^{-6}$. If however "*numerical noise*" is present in the computation of $f(\mathbf{x})$, special care should be taken in selecting δ_j. This may require doing some numerical experiments such as, for example, determining the sensitivity of approximation (2.6) to δ_j, for each j. Typically the sensitivity graph obtained is as depicted in Figure 2.7, and for the implementation of the optimization algorithm a value for δ_j should be chosen which corresponds to a point on the plateau as shown in Figure 2.7. Better approximations, at of course greater computational expense, may be obtained through the use of central finite differences.

2.3.2 Conjugate gradient methods

In spite of its local optimal descent property the method of steepest descent often performs poorly, following a zigzagging path of ever decreas-

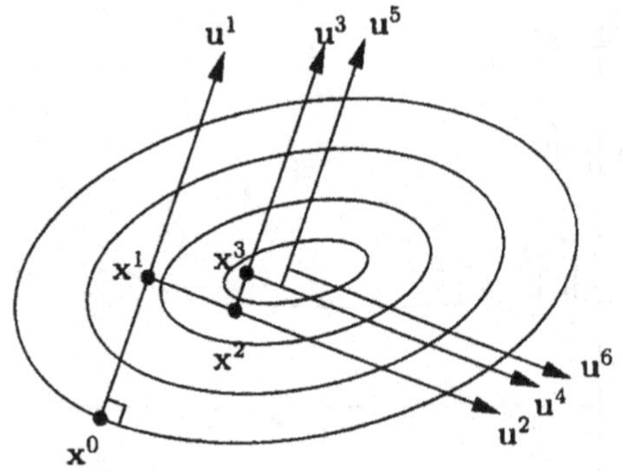

Figure 2.8: Orthogonal zigzagging behaviour of the steepest descent method

ing steps. This results in slow convergence and becomes extreme when the problem is poorly scaled, i.e. when the contours are extremely elongated. This poor performance is mainly due to the fact that the method enforces successive orthogonal search directions (see Section 2.3.1.3) as shown in Figure 2.8. Although, from a theoretical point of view, the method can be proved to be convergent, in practice the method may not effectively converge within a finite number of steps. Depending on the starting point this poor convergence also occurs when applying the method to positive-definite quadratic functions.

There is, however, a class of first order line search descent methods, known as *conjugate gradient methods*, for which it can be proved that whatever the scaling, a *method from this class will converge exactly in a finite number of iterations when applied to a positive-definite quadratic function*, i.e. to a function of the form

$$f(\mathbf{x}) = \tfrac{1}{2}\mathbf{x}^T\mathbf{A}\mathbf{x} + \mathbf{b}^T\mathbf{x} + c \tag{2.7}$$

where $c \in \mathbb{R}$, $\mathbf{b}$ is a real n-vector and $\mathbf{A}$ is a positive-definite $n \times n$ real symmetric matrix. Methods that have this property of *quadratic termination* are highly rated, because they are expected to also perform well on other non-quadratic functions in the neighbourhood of a local minimum. This is so, because by the Taylor expansion (1.21), it can be seen that many general differentiable functions approximate the form (2.7) near a local minimum.

2.3.2.1 Mutually conjugate directions

Two vectors $\mathbf{u}$, $\mathbf{v} \neq \mathbf{0}$ are defined to be *orthogonal* if the scalar product $\mathbf{u}^T\mathbf{v} = (\mathbf{u}, \mathbf{v}) = 0$. The concept of *mutual conjugacy* may be defined in a similar manner. Two vectors $\mathbf{u}$, $\mathbf{v} \neq \mathbf{0}$, are defined to be *mutually conjugate* with respect to the matrix $\mathbf{A}$ in (2.7) if $\mathbf{u}^T\mathbf{A}\mathbf{v} = (\mathbf{u}, \mathbf{A}\mathbf{v}) = 0$. Note that $\mathbf{A}$ is a positive definite symmetric matrix.

It can also be shown (see Theorem 6.5.1 in Chapter 6) that if the set of vectors $\mathbf{u}^i$, $i = 1, 2, \dots, n$ are *mutually conjugate*, then they form a *basis* in $\mathbb{R}^n$, i.e. any $\mathbf{x} \in \mathbb{R}^n$ may be expressed as

$$\mathbf{x} = \sum_{i=1}^{n} \lambda_i \mathbf{u}^i \tag{2.8}$$

where

$$\lambda_i = \frac{(\mathbf{u}^i, \mathbf{A}\mathbf{x})}{(\mathbf{u}^i, \mathbf{A}\mathbf{u}^i)}. \tag{2.9}$$

2.3.2.2 Convergence theorem for mutually conjugate directions

Suppose $\mathbf{u}^i$, $i = 1, 2, \dots, n$ are mutually conjugate with respect to positive-definite $\mathbf{A}$, then the optimal line search descent method in Section 2.1.1, using $\mathbf{u}^i$ as search directions, converges to the unique minimum $\mathbf{x}^*$ of $f(\mathbf{x}) = \frac{1}{2}\mathbf{x}^T\mathbf{A}\mathbf{x} + \mathbf{b}^T\mathbf{x} + c$ in less than or equal to n steps.

Proof:

If $\mathbf{x}^0$ the starting point, then after i iterations:

$$\begin{aligned}\mathbf{x}^i &= \mathbf{x}^{i-1} + \lambda_i \mathbf{u}^i = \mathbf{x}^{i-2} + \lambda_{i-1}\mathbf{u}^{i-1} + \lambda_i \mathbf{u}^i = \dots \\ &= \mathbf{x}^0 + \sum_{k=1}^{i} \lambda_k \mathbf{u}^k. \end{aligned} \tag{2.10}$$

The condition for *optimal descent* at iteration i is

$$\begin{aligned}\frac{dF(\lambda_i)}{d\lambda} = \left.\frac{df(\mathbf{x}^{i-1} + \lambda_i \mathbf{u}^i)}{d\lambda}\right|_{\mathbf{u}^i} &= [\mathbf{u}^i, \nabla f(\mathbf{x}^{i-1} + \lambda_i \mathbf{u}^i)] = 0 \\ &= [\mathbf{u}^i, \nabla f(\mathbf{x}^i)] = 0\end{aligned}$$

i.e.

$$\begin{aligned} 0 &= (\mathbf{u}^i, \mathbf{A}\mathbf{x}^i + \mathbf{b}) \\ &= \left(\mathbf{u}^i, \mathbf{A}\left(\mathbf{x}^0 + \sum_{k=1}^{i} \lambda_k \mathbf{u}^k\right) + \mathbf{b}\right) = (\mathbf{u}^i, \mathbf{A}\mathbf{x}^0 + \mathbf{b}) + \lambda_i(\mathbf{u}^i, \mathbf{A}\mathbf{u}^i) \end{aligned}$$

because $\mathbf{u}^i$, $i = 1, 2, \ldots, n$ are mutually conjugate, and thus

$$\lambda_i = -(\mathbf{u}^i, \mathbf{A}\mathbf{x}^0 + \mathbf{b})/(\mathbf{u}^i, \mathbf{A}\mathbf{u}^i). \tag{2.11}$$

Substituting (2.11) into (2.10) above gives

$$\begin{aligned} \mathbf{x}^n &= \mathbf{x}^0 + \sum_{i=1}^{n} \lambda_i \mathbf{u}^i = \mathbf{x}^0 - \sum_{i=1}^{n} \frac{(\mathbf{u}^i, \mathbf{A}\mathbf{x}^0 + \mathbf{b})\mathbf{u}^i}{(\mathbf{u}^i, \mathbf{A}\mathbf{u}^i)} \\ &= \mathbf{x}^0 - \sum_{i=1}^{n} \frac{(\mathbf{u}^i, \mathbf{A}\mathbf{x}^0)\mathbf{u}^i}{(\mathbf{u}^i, \mathbf{A}\mathbf{u}^i)} - \sum_{i=1}^{n} \frac{(\mathbf{u}^i, \mathbf{A}(\mathbf{A}^{-1}\mathbf{b}))\mathbf{u}^i}{(\mathbf{u}^i, \mathbf{A}\mathbf{u}^i)}. \end{aligned}$$

Now by utilizing (2.8) and (2.9) it follows that

$$\mathbf{x}^n = \mathbf{x}^0 - \mathbf{x}^0 - \mathbf{A}^{-1}\mathbf{b} = -\mathbf{A}^{-1}\mathbf{b} = \mathbf{x}^*.$$

The implication of the above theorem for the case $n = 2$ and where mutually conjugate line search directions $\mathbf{u}^1$ and $\mathbf{u}^2$ are used, is depicted in Figure 2.9.

2.3.2.3 Determination of mutually conjugate directions

How can mutually conjugate search directions be found? One way is to determine all the eigenvectors $\mathbf{u}^i$, $i = 1, 2, \ldots, n$ of $\mathbf{A}$. For $\mathbf{A}$ positive-definite, all the eigenvectors are mutually orthogonal and since $\mathbf{A}\mathbf{u}^i = \mu_i \mathbf{u}^i$ where μ_i is the associated eigenvalue, it follows directly that for all $i \neq j$ that $(\mathbf{u}^i, \mathbf{A}\mathbf{u}^j) = (\mathbf{u}^i, \mu_j \mathbf{u}^j) = \mu_j(\mathbf{u}^i, \mathbf{u}^j) = 0$, i.e. the eigenvectors are mutually conjugate with respect to $\mathbf{A}$. It is, however, not very practical to determine mutually conjugate directions by finding all the eigenvectors of $\mathbf{A}$, since the latter task in itself represents a computational problem of magnitude equal to that of solving the original unconstrained optimization problem via any other numerical algorithm. An easier method for obtaining mutually conjugate directions, is by means of the Fletcher-Reeves formulae (Fletcher and Reeves, 1964).

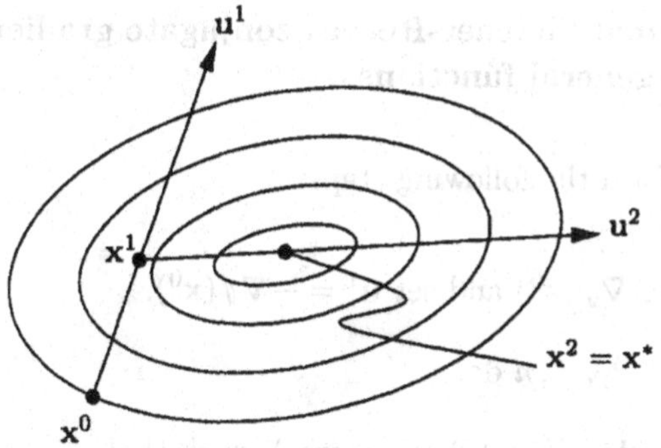

Figure 2.9: Quadratic termination of the conjugate gradient method in two steps for the case $n = 2$

2.3.2.4 The Fletcher-Reeves directions

The *Fletcher-Reeves directions* $\mathbf{u}^i$, $i = 1, 2, \dots, n$, that are listed below, can be shown (see Theorem 6.5.3 in Chapter 6) to be mutually conjugate with respect to the matrix $\mathbf{A}$ in the expression for the quadratic function in (2.7) (for which $\nabla f(\mathbf{x}) = \mathbf{A}\mathbf{x} + \mathbf{b}$). The explicit directions are:

$$\mathbf{u}^1 = -\nabla f(\mathbf{x}^0)$$

and for $i = 1, 2, \dots, n-1$

$$\mathbf{u}^{i+1} = -\nabla f(\mathbf{x}^i) + \beta_i \mathbf{u}^i \tag{2.12}$$

where $\mathbf{x}^i = \mathbf{x}^{i-1} + \lambda_i \mathbf{u}^i$, and λ_i corresponds to the optimal descent step in iteration i, and

$$\beta_i = \frac{\|\nabla f(\mathbf{x}^i)\|^2}{\|\nabla f(\mathbf{x}^{i-1})\|^2}. \tag{2.13}$$

The *Polak-Ribiere* directions are obtained if, instead of using (2.13), β_i is computed using

$$\beta_i = \frac{(\nabla f(\mathbf{x}^i) - \nabla f(\mathbf{x}^{i-1}))^T \nabla f(\mathbf{x}^i)}{\|\nabla f(\mathbf{x}^{i-1})\|^2}. \tag{2.14}$$

If $f(\mathbf{x})$ is quadratic it can be shown (Fletcher, 1987) that (2.14) is equivalent to (2.13).

2.3.2.5 Formal Fletcher-Reeves conjugate gradient algorithm for general functions

Given $\mathbf{x}^0$ perform the following steps:

1. Compute $\nabla f(\mathbf{x}^0)$ and set $\mathbf{u}^1 = -\nabla f(\mathbf{x}^0)$.
2. For $i = 1, 2, \ldots, n$ do:

   2.1 set $\mathbf{x}^i = \mathbf{x}^{i-1} + \lambda_i \mathbf{u}^i$ where λ_i such that

   $$f(\mathbf{x}^{i-1} + \lambda_i \mathbf{u}^i) = \min_{\lambda} f(\mathbf{x}^{i-1} + \lambda \mathbf{u}^i) \text{ (line search)},$$

   2.2 compute $\nabla f(\mathbf{x}^i)$,

   2.3 *if* convergence criteria satisfied, then STOP and $\mathbf{x}^* \cong \mathbf{x}^i$, *else* go to Step 2.4.

   2.4 *if* $1 \leq i \leq n-1$, $\mathbf{u}^{i+1} = -\nabla f(\mathbf{x}^i) + \beta_i \mathbf{u}^i$ with β_i given by (2.13).

3. Set $\mathbf{x}^0 = \mathbf{x}^n$ and go to Step 2 (restart).

If β_i is computed by (2.14) instead of (2.13) the method is known as the *Polak-Ribiere* method.

2.3.2.6 Simple illustrative example

Apply the Fletcher-Reeves method to minimize

$$f(\mathbf{x}) = \tfrac{1}{2}x_1^2 + x_1x_2 + x_2^2$$

with $\mathbf{x}^0 = [10, -5]^T$.

Solution:

Iteration 1:

$$\nabla f(\mathbf{x}) = \begin{bmatrix} x_1 + x_2 \\ x_1 + 2x_2 \end{bmatrix} \text{ and therefore } \mathbf{u}^1 = -\nabla f(\mathbf{x}^0) = \begin{bmatrix} -5 \\ 0 \end{bmatrix}.$$

$$\mathbf{x}^1 = \mathbf{x}^0 + \lambda \mathbf{u}^1 = \begin{bmatrix} 10 - 5\lambda \\ -5 \end{bmatrix} \text{ and}$$

$$F(\lambda) = f(\mathbf{x}^0 + \lambda \mathbf{u}^1) = \tfrac{1}{2}(10 - 5\lambda)^2 + (10 - 5\lambda)(-5) + 25.$$

For optimal descent

$$\frac{dF}{d\lambda}(\lambda) = \left.\frac{df}{d\lambda}\right|_{\mathbf{u}^1} = -5(10 - 5\lambda) + 25 = 0 \text{ (line search).}$$

This gives $\lambda_1 = 1$, $\mathbf{x}^1 = \begin{bmatrix} 5 \\ -5 \end{bmatrix}$ and $\nabla f(\mathbf{x}^1) = \begin{bmatrix} 0 \\ -5 \end{bmatrix}$.

Iteration 2:

$$\mathbf{u}^2 = -\nabla f(\mathbf{x}^1) + \frac{\|\nabla f(\mathbf{x}^1)\|^2}{\|\nabla f(\mathbf{x}^0)\|^2}\mathbf{u}^1 = -\begin{bmatrix} 0 \\ -5 \end{bmatrix} + \tfrac{25}{25}\begin{bmatrix} -5 \\ 0 \end{bmatrix} = \begin{bmatrix} -5 \\ 5 \end{bmatrix}.$$

$$\mathbf{x}^2 = \mathbf{x}^1 + \lambda \mathbf{u}^2 = \begin{bmatrix} 5 \\ -5 \end{bmatrix} + \lambda \begin{bmatrix} -5 \\ 5 \end{bmatrix} = \begin{bmatrix} 5(1-\lambda) \\ -5(1-\lambda) \end{bmatrix} \text{ and}$$

$$F(\lambda) = f(\mathbf{x}^1 + \lambda \mathbf{u}^2) = \tfrac{1}{2}[25(1-\lambda)^2 - 50(1-\lambda)^2 + 50(1-\lambda)^2].$$

Again for optimal descent

$$\frac{dF}{d\lambda}(\lambda) = \left.\frac{df}{d\lambda}\right|_{\mathbf{u}^2} = -25(1 - \lambda) = 0 \text{ (line search).}$$

This gives $\lambda_2 = 1$, $\mathbf{x}^2 = \begin{bmatrix} 0 \\ 0 \end{bmatrix}$ and $\nabla f(\mathbf{x}^2) = \begin{bmatrix} 0 \\ 0 \end{bmatrix}$. Therefore STOP.

The two iteration steps are shown in Figure 2.10.

2.4 Second order line search descent methods

These methods are based on Newton's method (see Section 1.5.4.1) for solving $\nabla f(\mathbf{x}) = \mathbf{0}$ iteratively: Given $\mathbf{x}^0$, then

$$\mathbf{x}^i = \mathbf{x}^{i-1} - \mathbf{H}^{-1}(\mathbf{x}^{i-1})\nabla f(\mathbf{x}^{i-1}),\ i = 1, 2, \ldots \tag{2.15}$$

As stated in Chapter 1, the main characteristics of this method are:

1. In the neighbourhood of the solution it may converge very fast, in

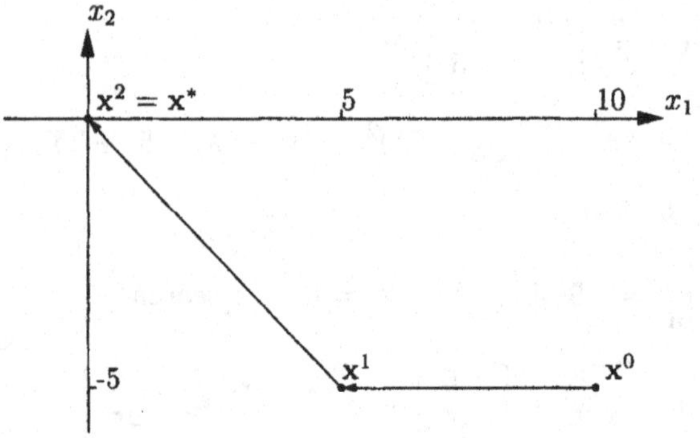

Figure 2.10: Convergence of Fletcher-Reeves method for illustrative example

fact it has the very desirous property of being quadratically convergent if it converges. Unfortunately convergence is not guaranteed and it may sometimes diverge, even from close to the solution.

2. The implementation of the method requires that $\mathbf{H}(\mathbf{x})$ be evaluated at each step.

3. To obtain the Newton step, $\boldsymbol{\Delta} = \mathbf{x}^i - \mathbf{x}^{i-1}$ it is also necessary to solve a $n \times n$ linear system $\mathbf{H}(\mathbf{x})\boldsymbol{\Delta} = -\boldsymbol{\nabla} f(\mathbf{x})$ at each iteration. This is computationally very expensive for large n, since an order n^3 multiplication operations are required to solve the system numerically.

2.4.1 Modified Newton's method

To avoid the problem of convergence (point 1. above), the computed Newton step $\boldsymbol{\Delta}$ is rather used as a search direction in the general line search descent algorithm given in Section 2.1.1. Thus at iteration i: select $\mathbf{u}^i = \boldsymbol{\Delta} = -\mathbf{H}^{-1}(\mathbf{x}^{i-1})\boldsymbol{\nabla} f(\mathbf{x}^{i-1})$, and minimize in that direction to obtain a λ_i such that

$$f(\mathbf{x}^{i-1} + \lambda_i \mathbf{u}^i) = \min_{\lambda} f(\mathbf{x}^{i-1} + \lambda \mathbf{u}^i)$$

and then set $\mathbf{x}^i = \mathbf{x}^{i-1} + \lambda_i \mathbf{u}^i$.

2.4.2 Quasi-Newton methods

To avoid the above mentioned computational problems (2. and 3.), methods have been developed in which approximations to $\mathbf{H}^{-1}$ are applied at each iteration. Starting with an approximation $\mathbf{G}_0$ to $\mathbf{H}^{-1}$ for the first iteration, the approximation is updated after each line search. An example of such a method is the Davidon-Fletcher-Powell (DFP) method.

2.4.2.1 DFP quasi-Newton method

The structure of this (rank-1 update) method (Fletcher, 1987) is as follows.

1. Choose $\mathbf{x}^0$ and set $\mathbf{G}_0 = \mathbf{I}$.

2. Do for iteration $i = 1, 2, \ldots, n$:

   2.1 set $\mathbf{x}^i = \mathbf{x}^{i-1} + \lambda_i \mathbf{u}^i$, where $\mathbf{u}^i = -\mathbf{G}_{i-1}\boldsymbol{\nabla} f(\mathbf{x}^{i-1})$ and λ_i is such that $f(\mathbf{x}^{i-1} + \lambda_i \mathbf{u}^i) = \min\limits_{\lambda} f(\mathbf{x}^{i-1} + \lambda \mathbf{u}^i)$, $\lambda_i \geq 0$ (line search),

   2.2 if stopping criteria satisfied then STOP, $\mathbf{x}^* \cong \mathbf{x}^i$,

   2.3 set $\mathbf{v}^i = \lambda_i \mathbf{u}^i$ and
   set $\mathbf{y}^i = \boldsymbol{\nabla} f(\mathbf{x}^i) - \boldsymbol{\nabla} f(\mathbf{x}^{i-1})$,

   2.4 set

$$\mathbf{G}_i = \mathbf{G}_{i-1} + \mathbf{A}_i + \mathbf{B}_i \text{ (rank 1-update)} \tag{2.16}$$

   where $\mathbf{A}_i = \dfrac{\mathbf{v}^i \mathbf{v}^{iT}}{\mathbf{v}^{iT}\mathbf{y}^i}$, $\mathbf{B}_i = \dfrac{-\mathbf{G}_{i-1}\mathbf{y}^i(\mathbf{G}_{i-1}\mathbf{y}^i)^T}{\mathbf{y}^{iT}\mathbf{G}_{i-1}\mathbf{y}^i}$.

3. Set $\mathbf{x}^0 = \mathbf{x}^n$; $\mathbf{G}_0 = \mathbf{G}_n$ (or $\mathbf{G}_0 = \mathbf{I}$), and go to Step 2 (restart).

2.4.2.2 Characteristics of DFP method

1. The method does not require the evaluation of $\mathbf{H}$ or the explicit solution of a linear system.

2. If $\mathbf{G}_{i-1}$ is positive-definite then so is $\mathbf{G}_i$ (see Theorem 6.6.1).

3. If $\mathbf{G}_i$ is positive-definite then descent is ensured at $\mathbf{x}^i$ because

$$\begin{aligned}\left.\frac{df(\mathbf{x}^i)}{d\lambda}\right|_{\mathbf{u}^{i+1}} &= \nabla^T f(\mathbf{x}^i)\mathbf{u}^{i+1}\\ &= -\nabla^T f(\mathbf{x}^i)\mathbf{G}_i\nabla f(\mathbf{x}^i) < 0, \text{ for all } \nabla f(\mathbf{x}) \neq \mathbf{0}.\end{aligned}$$

4. The directions $\mathbf{u}^i$, $i = 1, 2, \ldots, n$ are mutually conjugate for a quadratic function with $\mathbf{A}$ positive-definite (see Theorem 6.6.2). The method therefore possesses the desirable property of *quadratic termination* (see Section 2.3.2).

5. For quadratic functions: $\mathbf{G}_n = \mathbf{A}^{-1}$ (see again Theorem 6.6.2).

2.4.2.3 The BFGS method

The state-of-the-art quasi-Newton method is the Broyden-Fletcher-Goldfarb-Shanno (BFGS) method developed during the early 1970s (see Fletcher, 1987). This method uses a more complicated rank-2 update formula for $\mathbf{H}^{-1}$. For this method the update formula to be used in Step 2.4 of the algorithm given in Section 2.4.2.1 becomes

$$\begin{aligned}\mathbf{G}_i = \mathbf{G}_{i-1} &+ \left[1 + \frac{\mathbf{y}^{iT}\mathbf{G}_{i-1}\mathbf{y}^i}{\mathbf{v}^{iT}\mathbf{y}^i}\right]\left[\frac{\mathbf{v}^i\mathbf{v}^{iT}}{\mathbf{v}^{iT}\mathbf{y}^i}\right]\\ &- \left[\frac{\mathbf{v}^i\mathbf{y}^{iT}\mathbf{G}_{i-1} + \mathbf{G}_{i-1}\mathbf{y}^i\mathbf{v}^{iT}}{\mathbf{v}^{iT}\mathbf{y}^i}\right].\end{aligned} \tag{2.17}$$

2.5 Zero order methods and computer optimization subroutines

This chapter would not be complete without mentioning something about the large number of so-called *zero order* methods that have been developed. These methods are called such because they do not use either first order or second order derivative information, but only function values, i.e. only zero order derivative information.

Zero order methods are of the earliest methods and many of them are based on rough and ready ideas without very much theoretical background. Although these ad hoc methods are, as one may expect, much slower and computationally much more expensive than the higher order methods, they are usually reliable and easy to program. One of the most successful of these methods is the *simplex method* of Nelder and Mead (1965). This method should not be confused with the simplex method for linear programming. Another very powerful and popular method that only uses function values is the multi-variable method of Powell (1964). This method generates mutually conjugate directions by performing sequences of line searches in which only function evaluations are used. For this method Theorem 2.3.2.2 applies and the method therefore possesses the property if quadratic termination.

Amongst the more recently proposed and modern zero order methods, the method of *simulated annealing* and the so-called *genetic algorithms* (GAs) are the most prominent (see for example, Haftka and Gündel, 1992).

Computer programs are commercially available for all the unconstrained optimization methods presented in this chapter. Most of the algorithms may, for example, be found in the *Matlab Optimization Toolbox* and in the *IMSL* and *NAG* mathematical subroutine libraries.

2.6 Test functions

The efficiency of an algorithm is studied using standard functions with standard starting points $\mathbf{x}^0$. The total number of functions evaluations required to find $\mathbf{x}^*$ is usually taken as a measure of the efficiency of the algorithm. Some classical test functions (Rao, 1996) are listed below.

1. Rosenbrock's parabolic valley:

$$f(\mathbf{x}) = 100(x_2 - x_1^2)^2 + (1 - x_1)^2;\ \mathbf{x}^0 = \begin{bmatrix} -1.2 \\ 1.0 \end{bmatrix}\ \mathbf{x}^* = \begin{bmatrix} 1 \\ 1 \end{bmatrix}.$$

2. Quadratic function:

$$f(\mathbf{x}) = (x_1 + 2x_2 - 7)^2 + (2x_1 + x_2 - 5)^2;\ \mathbf{x}^0 = \begin{bmatrix} 0 \\ 0 \end{bmatrix}\ \mathbf{x}^* = \begin{bmatrix} 1 \\ 3 \end{bmatrix}.$$

3. Powell's quartic function:

$$\begin{aligned} f(\mathbf{x}) &= (x_1 + 10x_2)^2 + 5(x_3 - x_4)^2 + (x_2 - 2x_3)^4 + 10(x_1 - x_4)^4; \\ \mathbf{x}^0 &= [3, -1, 0, 1]^T; \ \mathbf{x}^* = [0, 0, 0, 0]^T. \end{aligned}$$

4. Fletcher and Powell's helical valley:

$$\begin{aligned} f(\mathbf{x}) &= 100\left((x_3 - 10\theta(x_1, x_2))^2 + \left(\sqrt{x_1^2 + x_2^2} - 1\right)^2\right) + x_3^2; \\ \text{where } 2\pi\theta(x_1, x_2) &= \begin{cases} \arctan \dfrac{x_2}{x_1} & \text{if } x_1 > 0 \\ \pi + \arctan \dfrac{x_2}{x_1} & \text{if } x_1 < 0 \end{cases} \\ \mathbf{x}^0 &= [-1, 0, 0]^T; \ \mathbf{x}^* = [1, 0, 0]^T. \end{aligned}$$

5. A non-linear function of three variables:

$$\begin{aligned} f(\mathbf{x}) &= \frac{1}{1 + (x_1 - x_2)^2} + \sin\left(\frac{1}{2}\pi x_2 x_3\right) + \exp\left(-\left(\frac{x_1 + x_3}{x_2} - 2\right)^2\right); \\ \mathbf{x}^0 &= [0, 1, 2]^T; \ \mathbf{x}^* = [1, 1, 1]^T. \end{aligned}$$

6. Freudenstein and Roth function:

$$\begin{aligned} f(\mathbf{x}) &= (-13 + x_1 + ((5 - x_2)x_2 - 2)x_2)^2 \\ &\quad + (-29 + x_1 + ((x_2 + 1)x_2 - 14)x_2)^2; \\ \mathbf{x}^0 &= [0.5, -2]^T; \ \mathbf{x}^* = [5, 4]^T; \ \mathbf{x}^*_{\text{local}} = [11.41\ldots, -0.8968\ldots]^T. \end{aligned}$$

7. Powell's badly scaled function:

$$\begin{aligned} f(\mathbf{x}) &= (10\,000 x_1 x_2 - 1)^2 + (\exp(-x_1) + \exp(-x_2) - 1.0001)^2; \\ \mathbf{x}^0 &= [0, 1]^T; \ \mathbf{x}^* = [1.098 \cdots \times 10^{-5}, 9.106\ldots]^T. \end{aligned}$$

8. Brown's badly scaled function:

$$\begin{aligned} f(\mathbf{x}) &= (x_1 - 10^6)^2 + (x_2 - 2 \times 10^{-6})^2 + (x_1 x_2 - 2)^2; \\ \mathbf{x}^0 &= [1, 1]^T; \ \mathbf{x}^* = [10^6, 2 \times 10^6]^T. \end{aligned}$$

9. Beale's function:

$$\begin{aligned} f(\mathbf{x}) &= (1.5 - x_1(1 - x_2))^2 + (2.25 - x_1(1 - x_2^2))^2 \\ &\quad +(2.625 - x_1(1 - x_2^3))^2; \\ \mathbf{x}^0 &= [1, 1]^T; \;\; \mathbf{x}^* = [3, 0.5]^T. \end{aligned}$$

10. Wood's function:

$$\begin{aligned} f(\mathbf{x}) &= 100(x_2 - x_1^2)^2 + (1 - x_1)^2 + 90(x_4 - x_3^2)^2 + (1 - x_3)^2 \\ &\quad +10(x_2 + x_4 - 2)^2 + 0.1(x_2 - x_4)^2 \\ \mathbf{x}^0 &= [-3, -1, -3, -1]^T; \;\; \mathbf{x}^* = [1, 1, 1, 1]^T. \end{aligned}$$

Chapter 3

STANDARD METHODS FOR CONSTRAINED OPTIMIZATION

3.1 Penalty function methods for constrained minimization

Consider the general constrained optimization problem:

$$\begin{array}{lll} \underset{\mathbf{x}}{\text{minimize}} & f(\mathbf{x}) & \\ \text{such that} & g_j(\mathbf{x}) \leq 0 \;\; j = 1, 2, \ldots, m & \\ & h_j(\mathbf{x}) = 0 \;\; j = 1, 2, \ldots, r. & \end{array} \tag{3.1}$$

The most simple and straight forward approach to handling constrained problems of the above form is to apply a suitable unconstrained optimization algorithm to *a penalty function formulation of* constrained problem (3.1).

3.1.1 The penalty function formulation

The penalty function formulation of the general constrained problem (3.1) is

$$\underset{\mathbf{x}}{\text{minimize}}\, P(\mathbf{x})$$

where

$$P(\mathbf{x},\boldsymbol{\rho},\boldsymbol{\beta}) = f(\mathbf{x}) + \sum_{j=1}^{r} \rho_j h_j^2(\mathbf{x}) + \sum_{j=1}^{m} \beta_j g_j^2(\mathbf{x}) \tag{3.2}$$

and where the components of the penalty parameter vectors $\boldsymbol{\rho}$ and $\boldsymbol{\beta}$ are given by

$$\rho_j \gg 0;\ \beta_j = \begin{cases} 0 & \text{if } g_j(\mathbf{x}) \leq 0 \\ \mu_j \gg 0 & \text{if } g_j(\mathbf{x}) > 0. \end{cases}$$

The latter parameters, ρ_j and β_j, are called *penalty parameters*, and $P(\mathbf{x},\boldsymbol{\rho},\boldsymbol{\beta})$ the *penalty function*. The solution to this unconstrained minimization problem is denoted by $\mathbf{x}^*(\boldsymbol{\rho},\boldsymbol{\beta})$, where $\boldsymbol{\rho}$ and $\boldsymbol{\beta}$ denote the respective vectors of penalty parameters.

Often $\rho_j \equiv$ constant $\equiv \rho$, for all j, and also $\mu_j \equiv \rho$ for all j such that $g_j(\mathbf{x}) > 0$. Thus P in (3.2) is denoted by $P(\mathbf{x},\rho)$ and the corresponding minimum by $\mathbf{x}^*(\rho)$. It can be shown that under normal continuity conditions the $\lim_{\rho\to\infty} \mathbf{x}^*(\rho) = \mathbf{x}^*$. Typically the overall penalty parameter ρ is set at $\rho = 10^4$ if the constraints functions are normalized in some sense.

3.1.2 Illustrative examples

Consider the following two one-dimensional constrained optimization problems:

(a)

$$\begin{aligned} & \min f(x) \\ \text{such that}\quad & h(x) = x - a = 0, \\ \text{then}\quad & P(x) = f(x) + \rho(x-a)^2; \end{aligned}$$

and

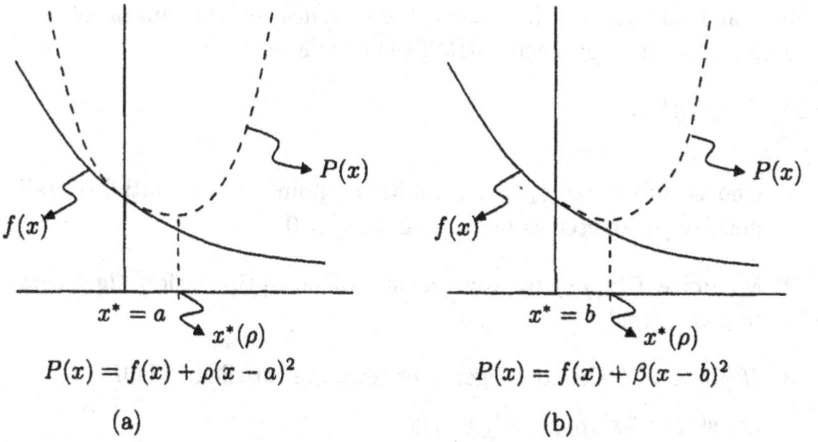

Figure 3.1: Behaviour of penalty function for one-dimensional (a) equality constrained, and (b) inequality constrained minimization problems

(b)

$$\begin{aligned} & \min f(x) \\ \text{such that} \quad & g(x) = x - b \leq 0, \\ \text{then} \quad & P(x) = f(x) + \beta(x - b)^2. \end{aligned}$$

The penalty function solutions to these two problems are as depicted in Figures 3.1 (a) and (b). The penalty function method falls in the class of *external methods* because it converges externally from the infeasible region.

3.1.3 Sequential unconstrained minimization technique (SUMT)

Unfortunately the penalty function method becomes unstable and inefficient for very large ρ if high accuracy is required. This is because rounding errors result in the computation of unreliable descent directions. If second order unconstrained minimization methods are used for the minimization of $P(\mathbf{x}, \rho)$, then the associated Hessian matrices become ill-conditioned and again the method is inclined to break down. A remedy to this situation is to apply the penalty function method to a sequence of sub-problems, starting with moderate penalty parameter values, and successively increasing their values for the sub-problems. The details of this approach (SUMT) is as follows.

SUMT algorithm:

1. Choose tolerances ε_1 and ε_2, starting point $\mathbf{x}^0$ and initial overall penalty parameter value ρ_0, and set $k = 0$.
2. Minimize $P(\mathbf{x}, \rho_k)$ by any unconstrained optimization algorithm to give $\mathbf{x}^*(\rho_k)$.
3. *If* (for $k > 0$) the convergence criteria are satisfied: STOP

   i.e. stop if $\|\mathbf{x}^*(\rho_k) - \mathbf{x}^*(\rho_{k-1})\| < \varepsilon_1$

   and/or $|P(\mathbf{x}^*(\rho_{k-1}) - P(\mathbf{x}^*(\rho_k))| < \varepsilon_2$,

   else

   set $\rho_{k+1} = c\rho_k$, $c > 1$ and $\mathbf{x}^0 = \mathbf{x}^*(\rho_k)$,

   set $k = k + 1$ and go to Step 2.

Typically choose $\rho_0 = 1$ and $c = 10$.

3.1.4 Simple example

Consider the constrained problem:

$$\min f(\mathbf{x}) = \tfrac{1}{3}(x_1 + 1)^3 + x_2$$

such that

$$1 - x_1 \leq 0; \ -x_2 \leq 0.$$

The problem is depicted in Figure 3.2.

Define the penalty function:

$$P(\mathbf{x}, \rho) = \tfrac{1}{3}(x_1 + 1)^3 + x_2 + \rho(1 - x_1)^2 + \rho x_2^2.$$

(Of course the ρ only comes into play if the corresponding constraint is violated.) The penalty function solution may now be obtained analytically as follows.

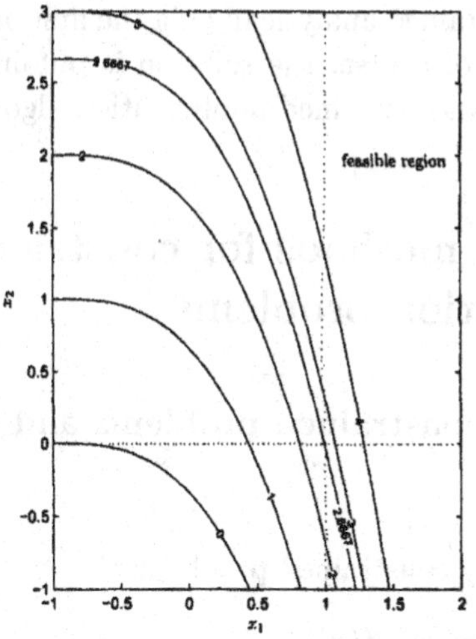

Figure 3.2: Contour representation of objective function and feasible region for simple example

The first order necessary conditions for an unconstrained minimum of P are

$$\frac{\partial P}{\partial x_1} = (x_1 + 1)^2 - 2\rho(1 - x_1) = 0 \tag{3.3}$$

$$\frac{\partial P}{\partial x_2} = 1 + 2\rho x_2 = 0. \tag{3.4}$$

From (3.4): $x_2^*(\rho) = -\frac{1}{2\rho}$

and from (3.3): $x_1^2 + 2(1+\rho)x_1 + 1 - 2\rho = 0$.

Solving the quadratic equation and taking the positive root gives

$$x_1^*(\rho) = -(1+\rho) + (1+\rho)\left(1 + \frac{(2\rho - 1)}{(1+\rho)^2}\right)^{1/2}.$$

Clearly $\lim_{\rho\to\infty} x_2^*(\rho) = 0$ and $\lim_{\rho\to\infty} x_1^*(\rho) = 1$ (where use has been made of the expansion $(1+\varepsilon)^{1/2} = 1 + \frac{1}{2}\varepsilon + \dots$) giving $\mathbf{x}^* = [1, 0]^T$, as one expects from Figure 3.2. Here the solution to the penalty function formulated problem has been obtained analytically via the first order necessary conditions. In general, of course, the solution is obtained numerically by applying a suitable unconstrained minimization algorithm.

3.2 Classical methods for constrained optimization problems

3.2.1 Equality constrained problems and the Lagrangian function

Consider the equality constrained problem:

$$\begin{array}{ll} \text{minimize} & f(\mathbf{x}) \\ \text{such that} & h_j(\mathbf{x}) = 0, \ j = 1, 2, \dots, r < n. \end{array} \tag{3.5}$$

In 1760 Lagrange transformed this constrained problem to an unconstrained problem via the introduction of so-called *Lagrange multipliers* λ_j, $j = 1, 2, \dots, r$ in the formulation of the *Lagrangian function:*

$$L(\mathbf{x}, \boldsymbol{\lambda}) = f(\mathbf{x}) + \sum_{j=1}^{r} \lambda_j h_j(\mathbf{x}) = f(\mathbf{x}) + \boldsymbol{\lambda}^T \mathbf{h}(\mathbf{x}). \tag{3.6}$$

The necessary conditions for a constrained minimum of the above equality constrained problem may be stated in terms of the Lagrangian function and the Lagrange multipliers.

3.2.1.1 Necessary conditions for an equality constrained minimum

Let the functions f and $h_j \in C^1$ then, on the assumption that the $n \times r$ *Jacobian* matrix

$$\frac{\partial \mathbf{h}(\mathbf{x}^*)}{\partial \mathbf{x}} = [\boldsymbol{\nabla} h_1(\mathbf{x}^*), \boldsymbol{\nabla} h_2(\mathbf{x}^*), \dots]$$

is of rank r, the *necessary conditions* for $\mathbf{x}^*$ to be a constrained internal *local minimum* of the equality constrained problem (3.5) is that $\mathbf{x}^*$ corresponds to a stationary point $(\mathbf{x}^*, \boldsymbol{\lambda}^*)$ of the Lagrangian function, i.e.

that a vector $\boldsymbol{\lambda}^*$ exists such that

$$\begin{aligned}\frac{\partial L}{\partial x_i}(\mathbf{x}^*, \boldsymbol{\lambda}^*) &= 0, \ i = 1, 2, \dots, n \\ \frac{\partial L}{\partial \lambda_j}(\mathbf{x}^*, \boldsymbol{\lambda}^*) &= 0, \ j = 1, 2, \dots, r.\end{aligned} \tag{3.7}$$

For a formal proof of the above, see Theorem 6.2.1.

3.2.1.2 The Lagrangian method

Note that necessary conditions (3.7) represent $n+r$ equations in the $n+r$ unknowns $x_1^*, x_2^*, \dots, x_n^*, \lambda_1^*, \dots, \lambda_r^*$. The solutions to these, in general non-linear equations, therefore give candidate solutions $\mathbf{x}^*$ to problem (3.5). This indirect approach to solving the constrained problem is illustrated in solving the following simple example problem.

3.2.1.3 Example

$$\begin{aligned}\text{minimize} \quad & f(\mathbf{x}) = (x_1 - 2)^2 + (x_2 - 2)^2 \\ \text{such that} \quad & h(\mathbf{x}) = x_1 + x_2 - 6 = 0.\end{aligned}$$

First formulate the Lagrangian:

$$L(\mathbf{x}, \lambda) = (x_1 - 2)^2 + (x_2 - 2)^2 + \lambda(x_1 + x_2 - 6).$$

By the Theorem in Section 3.2.1.1 the necessary conditions for a constrained minimum are

$$\begin{aligned}\frac{\partial L}{\partial x_1} &= 2(x_1 - 2) + \lambda = 0 \\ \frac{\partial L}{\partial x_2} &= 2(x_2 - 2) + \lambda = 0 \\ \frac{\partial L}{\partial \lambda} &= x_1 + x_2 - 6 = 0.\end{aligned}$$

Solving these equations gives a candidate point: $x_1^* = 3$, $x_2^* = 3$, $\lambda^* = -2$ with $f(\mathbf{x}^*) = 2$. This solution is depicted in Figure 3.3

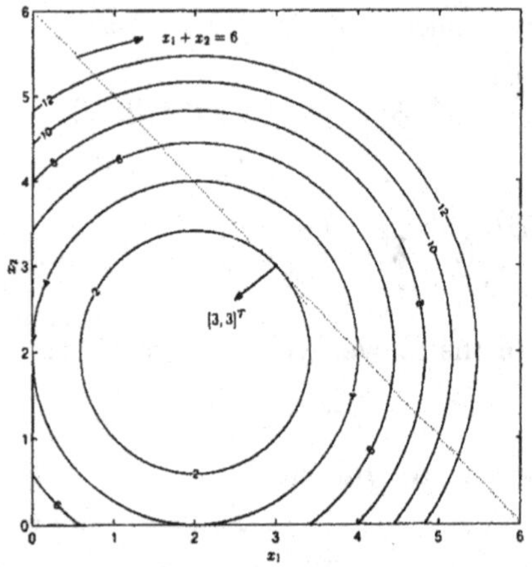

Figure 3.3: Graphical solution to example problem 3.2.1.3

3.2.1.4 Sufficient conditions

In general the necessary conditions (3.7) are not sufficient to imply a constrained local minimum at $\mathbf{x}^*$. A more general treatment of the sufficiency conditions is however very complicated and will not be discussed here. It can, however, be shown that if $f(\mathbf{x})$ and the $h_j(\mathbf{x})$ are all convex functions, with $\lambda_j^* > 0$ for all j, then conditions (3.7) indeed constitute sufficiency conditions. In this case the local constrained minimum is unique and represents the global minimum.

3.2.1.5 Saddle point of the Lagrangian function

Assume

(i) $f(\mathbf{x})$ has a constrained minimum at $\mathbf{x}^*$ (with associated $\boldsymbol{\lambda}^*$) and

(ii) that if $\boldsymbol{\lambda}$ is chosen in the neighbourhood of $\boldsymbol{\lambda}^*$, then $L(\mathbf{x}, \boldsymbol{\lambda})$ has a local minimum with respect to $\mathbf{x}$ in the neighbourhood of $\mathbf{x}^*$. (This can be expected if the Hessian matrix of L with respect to $\mathbf{x}$ at $(\mathbf{x}^*, \boldsymbol{\lambda}^*)$ is positive definite.)

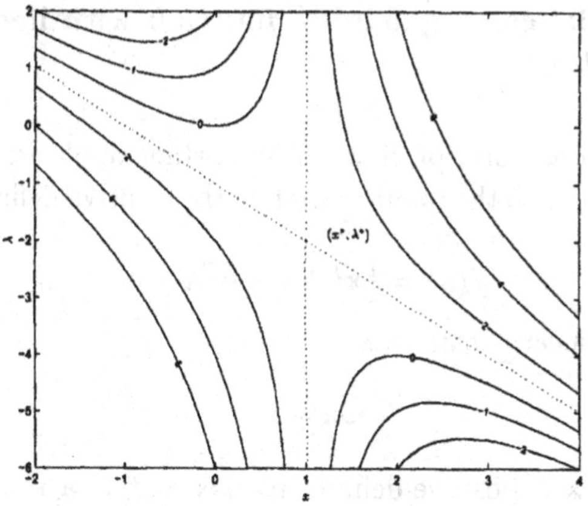

Figure 3.4: Saddle point of Lagrangian function $L = x^2 + \lambda(x-1)$

It *can be shown* (see Theorem 6.2.2) that if (i) and (ii) applies then $L(\mathbf{x}, \boldsymbol{\lambda})$ has a saddle point at $(\mathbf{x}^*, \boldsymbol{\lambda}^*)$. Indeed it is a degenerate saddle point since

$$L(\mathbf{x}, \boldsymbol{\lambda}^*) \geq L(\mathbf{x}^*, \boldsymbol{\lambda}^*) = L(\mathbf{x}^*, \boldsymbol{\lambda}).$$

Consider the example:

$$\text{minimize } f(x) = x^2 \text{ such that } h(x) = x - 1 = 0.$$

With $L = x^2 + \lambda(x-1)$ it follows directly that $x^* = 1$ and $\lambda^* = -2$.

Since along the straight line asymptotes through (x^*, λ^*), $\Delta L = 0$ for changes Δx and $\Delta\lambda$, it follows that

$$[\Delta x \; \Delta\lambda] \begin{bmatrix} 2 & 1 \\ 1 & 0 \end{bmatrix} \begin{bmatrix} \Delta x \\ \Delta\lambda \end{bmatrix} = 0, \text{ or}$$

$$\Delta x(\Delta x + \Delta\lambda) = 0.$$

The asymptotes therefore are the lines through (x^*, λ^*) with $\Delta x = 0$, and $\frac{\Delta\lambda}{\Delta x} = -1$ respectively, as shown in Figure 3.4, i.e. the lines $x = 1$ and $\lambda = -x - 1$.

In general, if it can be shown that candidate point $(\mathbf{x}^*, \boldsymbol{\lambda}^*)$ is a saddle point of L, then the Hessian of the Lagrangian with respect to $\mathbf{x}$, $\mathbf{H}_L$, at the saddle point is positive definite for a constrained minimum.

3.2.1.6 Special case: quadratic function with linear equality constraints

From a theoretical point of view, an important application of the Lagrangian method is to the minimization of the positive-definite quadratic function:

$$f(\mathbf{x}) = \tfrac{1}{2}\mathbf{x}^T\mathbf{A}\mathbf{x} + \mathbf{b}^T\mathbf{x} + c \tag{3.8}$$

subject to the linear constraints

$$\mathbf{C}\mathbf{x} = \mathbf{d}.$$

Here $\mathbf{A}$ is a $n \times n$ positive-definite matrix and $\mathbf{C}$ a $r \times n$ constraint matrix, $r < n$, $\mathbf{b}$ is a n-vector and $\mathbf{d}$ a r-vector.

In this case the Lagrangian is

$$L(\mathbf{x}, \boldsymbol{\lambda}) = \tfrac{1}{2}\mathbf{x}^T\mathbf{A}\mathbf{x} + \mathbf{b}^T\mathbf{x} + c + \boldsymbol{\lambda}^T(\mathbf{C}\mathbf{x} - \mathbf{d})$$

and the necessary conditions (3.7) for a constrained minimum at $\mathbf{x}^*$ is the existence of a vector $\boldsymbol{\lambda}^*$ such that

$$\begin{aligned}
\nabla_{\mathbf{x}} L(\mathbf{x}^*, \boldsymbol{\lambda}^*) &= \mathbf{A}\mathbf{x}^* + \mathbf{b} + \mathbf{C}^T\boldsymbol{\lambda}^* = \mathbf{0}\\
\nabla_{\boldsymbol{\lambda}} L(\mathbf{x}^*, \boldsymbol{\lambda}^*) &= \mathbf{C}\mathbf{x}^* - \mathbf{d} = \mathbf{0}
\end{aligned}$$

i.e.

$$\begin{bmatrix} \mathbf{A} & \mathbf{C}^T \\ \mathbf{C} & \mathbf{0} \end{bmatrix} \begin{bmatrix} \mathbf{x}^* \\ \boldsymbol{\lambda}^* \end{bmatrix} = \begin{bmatrix} -\mathbf{b} \\ \mathbf{d} \end{bmatrix}. \tag{3.9}$$

The solution to this linear system is given by

$$\begin{bmatrix} \mathbf{x}^* \\ \boldsymbol{\lambda}^* \end{bmatrix} = \mathbf{M}^{-1} \begin{bmatrix} -\mathbf{b} \\ \mathbf{d} \end{bmatrix} \text{ where } \mathbf{M} = \begin{bmatrix} \mathbf{A} & \mathbf{C}^T \\ \mathbf{C} & \mathbf{0} \end{bmatrix}.$$

3.2.1.7 Inequality constraints as equality constraints

Consider the more general problem:

$$\begin{aligned}
&\text{minimize} && f(\mathbf{x})\\
&\text{such that} && g_j(\mathbf{x}) \leq 0,\ j = 1, 2, \ldots, m\\
& && h_j(\mathbf{x}) = 0,\ j = 1, 2, \ldots, r.
\end{aligned} \tag{3.10}$$

The inequality constraints may be transformed to equality constraints by the introduction of so-called *auxiliary variables* θ_j, $j = 1, 2, \ldots, m$:

$$g_j(\mathbf{x}) + \theta_j^2 = 0.$$

Since $g_j(\mathbf{x}) = -\theta_j^2 \leq 0$ for all j, the inequality constraints are automatically satisfied.

The Lagrangian method for equality constrained problems may now be applied, where

$$L(\mathbf{x}, \boldsymbol{\theta}, \boldsymbol{\lambda}, \boldsymbol{\mu}) = f(\mathbf{x}) + \sum_{j=1}^{m} \lambda_j (g_j(\mathbf{x}) + \theta_j^2) + \sum_{j=1}^{r} \mu_j h_j(\mathbf{x}) \tag{3.11}$$

and λ_j and μ_j denote the respective Lagrange multipliers.

From (3.7) the associated necessary conditions for a minimum at $\mathbf{x}$ are

$$\begin{aligned}
\frac{\partial L}{\partial x_i} &= \frac{\partial f(\mathbf{x})}{\partial x_i} + \sum_{j=1}^{m} \lambda_j \frac{\partial g_j(\mathbf{x})}{\partial x_i} + \sum_{j=1}^{r} \mu_j \frac{\partial h_j(\mathbf{x})}{\partial x_i} = 0, \ i = 1, 2, \ldots, n \\
\frac{\partial L}{\partial \theta_j} &= 2\lambda_j \theta_j = 0, \ j = 1, 2, \ldots, m \\
\frac{\partial L}{\partial \lambda_j} &= g_j(\mathbf{x}) + \theta_j^2 = 0, \ j = 1, 2, \ldots, m \\
\frac{\partial L}{\partial \mu_j} &= h_j(\mathbf{x}) = 0, \ j = 1, 2, \ldots, r.
\end{aligned} \tag{3.12}$$

The above system (3.12) represents a system of $n + 2m + r$ simultaneous non-linear equations in the $n + 2m + r$ unknowns $\mathbf{x}$, $\boldsymbol{\theta}$, $\boldsymbol{\lambda}$ and $\boldsymbol{\mu}$. Obtaining the solutions to system (3.12) yields candidate solutions $\mathbf{x}^*$ to the general optimization problem (3.10). The application of this approach is demonstrated in the following example.

3.2.1.8 Example

Minimize $f(\mathbf{x}) = 2x_1^2 - 3x_2^2 - 2x_1$

such that $x_1^2 + x_2^2 \leq 1$ by making use of auxiliary variables.

Introduce auxiliary variable θ such that

$$x_1^2 + x_2^2 - 1 + \theta^2 = 0$$

then

$$L(\mathbf{x}, \theta, \lambda) = 2x_1^2 - 3x_2^2 - 2x_1 + \lambda(x_1^2 + x_2^2 - 1 + \theta^2).$$

The necessary conditions at the minimum are

$$\frac{\partial L}{\partial x_1} = 4x_1 - 2 + 2\lambda x_1 = 0 \quad (3.13)$$

$$\frac{\partial L}{\partial x_2} = -6x_2 + 2\lambda x_2 = 0 \quad (3.14)$$

$$\frac{\partial L}{\partial \theta} = 2\lambda\theta = 0 \quad (3.15)$$

$$\frac{\partial L}{\partial \lambda} = x_1^2 + x_2^2 - 1 + \theta^2 = 0. \quad (3.16)$$

As first choice in (3.15), select $\lambda = 0$ which gives $x_1 = 1/2$, $x_2 = 0$, and $\theta^2 = 3/4$.

Since $\theta^2 > 0$ the problem is unconstrained at this point. Further since $\mathbf{H} = \begin{bmatrix} 4 & 0 \\ 0 & -6 \end{bmatrix}$ is non-definite the candidate point $\mathbf{x}^0$ corresponds to a saddle point where $f(\mathbf{x}^0) = -0.5$.

Select as second choice in (3.15),

$$\theta = 0 \text{ which gives } x_1^2 + x_2^2 - 1 = 0 \quad (3.17)$$

i.e. the constraint is active. From (3.14) it follows that for $x_2 \neq 0$, $\lambda = 3$, and substituting into (3.13) gives $x_1 = 1/5$ and from (3.17): $x_2 = \pm\sqrt{24}/5 = \pm 0.978$ which give the two possibilities: $\mathbf{x}^* = \left(\frac{1}{5}, \frac{\sqrt{24}}{5}\right)$, $f(\mathbf{x}^*) = -3.189$ and $\mathbf{x}^* = \left(\frac{1}{5}, \frac{-\sqrt{24}}{5}\right)$, with $f(\mathbf{x}^*) = -3.189$.

3.2.1.9 Direction of asymptotes at a saddle point $\mathbf{x}^0$

If, in the example above, the direction of an asymptote at saddle point $\mathbf{x}^0$ is denoted by the unit vector $\mathbf{u} = [u_1, u_2]^T$, then for a displacement

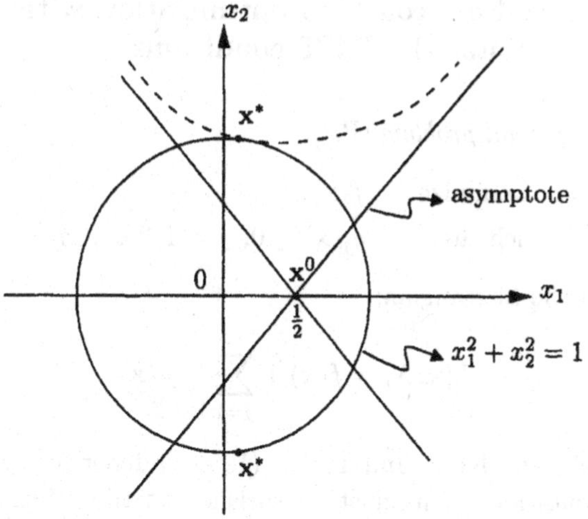

Figure 3.5: Directions of asymptotes at a saddle point $\mathbf{x}^0$

$\Delta\mathbf{x} = \mathbf{u}$ along the asymptote the change in the function value is $\Delta f = 0$. It follows from the Taylor expansion that

$$\Delta f = f(\mathbf{x}^0 + \Delta\mathbf{x}) - f(\mathbf{x}^0) = \mathbf{u}^T \nabla f(\mathbf{x}^0) + \tfrac{1}{2}\mathbf{u}^T \mathbf{H}\mathbf{u} = 0$$

for step $\Delta\mathbf{x} = \mathbf{u}$ at saddle point $\mathbf{x}^0$. Since $\mathbf{H} = \begin{bmatrix} 4 & 0 \\ 0 & -6 \end{bmatrix}$ and $\nabla f(\mathbf{x}^0) = \mathbf{0}$ it follows that $2u_1^2 - 3u_2^2 = 0$ and also, since $\|\mathbf{u}\| = 1$: $u_1^2 + u_2^2 = 1$.

Solving for u_1 and u_2 in the above gives

$$u_1 = \pm\sqrt{\tfrac{3}{5}};\ u_2 = \pm\sqrt{\tfrac{2}{5}}$$

which, taken in combinations, correspond to the directions of the four asymptotes at the saddle point $\mathbf{x}^0$, as shown in Figure 3.5.

3.2.2 Classical approach to optimization with inequality constraints: the KKT conditions

Consider the *primal problem* (PP):

$$\begin{array}{ll} \text{minimize} & f(\mathbf{x}) \\ \text{such that} & g_j(\mathbf{x}) \leq 0,\ j = 1, 2, \dots, m. \end{array} \tag{3.18}$$

Define again the Lagrangian:

$$L(\mathbf{x}, \boldsymbol{\lambda}) = f(\mathbf{x}) + \sum_{j=1}^{m} \lambda_j g_j(\mathbf{x}). \tag{3.19}$$

Karush (1939) and Kuhn and Tucker (1951) independently derived the necessary conditions that must be satisfied at the solution $\mathbf{x}^*$ of the primary problem (3.18). These conditions are generally known as the *KKT conditions* which are expressed in terms of the Lagrangian $L(\mathbf{x}, \boldsymbol{\lambda})$.

3.2.2.1 The KKT necessary conditions for an inequality constrained minimum

Let the functions f and $g_j \in C^1$, and assume the existence of Lagrange multipliers $\boldsymbol{\lambda}^*$, then at the point $\mathbf{x}^*$, corresponding to the solution of the primal problem (3.18), the following conditions must be satisfied:

$$\begin{aligned} \frac{\partial f}{\partial x_i}(\mathbf{x}^*) + \sum_{j=1}^{m} \lambda_j^* \frac{\partial g_j}{\partial x_i}(\mathbf{x}^*) &= 0,\ i = 1, 2, \dots, n \\ g_j(\mathbf{x}^*) &\leq 0,\ j = 1, 2, \dots, m \\ \lambda_j^* g_j(\mathbf{x}^*) &= 0,\ j = 1, 2, \dots, m \\ \lambda_j^* &\geq 0,\ j = 1, 2, \dots, m. \end{aligned} \tag{3.20}$$

For a formal proof of the above see Theorem 6.3.1. It can be shown that the KKT conditions also constitute *sufficient* conditions (those that imply that) for $\mathbf{x}^*$ to be a constrained minimum, if $f(\mathbf{x})$ and the $g_j(\mathbf{x})$ are all convex functions.

Let $\mathbf{x}^*$ be a solution to problem (3.18), and suppose that the KKT conditions (3.20) are satisfied. If now $g_k(\mathbf{x}^*) = 0$ for some $k \in \{1, 2, \dots, m\}$, then the corresponding inequality constraint k is said to be *active* and *binding* at $\mathbf{x}^*$, if the corresponding Lagrange multiplier $\lambda_k^* \geq 0$. It is strongly active if $\lambda_k^* > 0$, and weakly active if $\lambda_k^* = 0$. However, if for some candidate KKT point $\bar{\mathbf{x}}$, $g_k(\bar{\mathbf{x}}) = 0$ for some k, and all the KKT conditions are satisfied except that the corresponding Lagrange multiplier $\bar{\lambda}_k < 0$, then the inequality constraint k is said to be inactive, and must be deleted form the set of active constraints at $\bar{\mathbf{x}}$.

3.2.2.2 Constraint qualification

It can be shown that the existence of $\boldsymbol{\lambda}^*$ is guaranteed if the so-called *constraint qualification* is satisfied at $\mathbf{x}^*$, i.e. if a vector $\mathbf{h} \in \mathbb{R}^n$ exists such that for each active constraint j at $\mathbf{x}^*$

$$\boldsymbol{\nabla}^T g_j(\mathbf{x}^*)\mathbf{h} < 0 \tag{3.21}$$

then $\boldsymbol{\lambda}^*$ exists.

The constraint qualification (3.21) is always satisfied

(i) if all the constraints are convex and at least one $\mathbf{x}$ exists within the feasible region, or

(ii) if the rank of the Jacobian of all active and binding constraints at $\mathbf{x}^*$ is maximal, or

(iii) if all the constraints are linear.

3.2.2.3 Illustrative example

Minimize $f(\mathbf{x}) = (x_1 - 2)^2 + x_2^2$

such that $x_1 \geq 0$, $x_2 \geq 0$, $(1 - x_1)^3 \geq x_2$.

Clearly minimum exists at $x_1^* = 1$, $x_2^* = 0$ where

$$g_1(\mathbf{x}^*) = -x_1^* < 0,\ g_2(\mathbf{x}^*) = -x_2^* = 0 \text{ and } g_3(\mathbf{x}^*) = x_2^* - (1 - x_1^*)^3 = 0$$

and therefore g_2 and g_3 are active at $\mathbf{x}^*$.

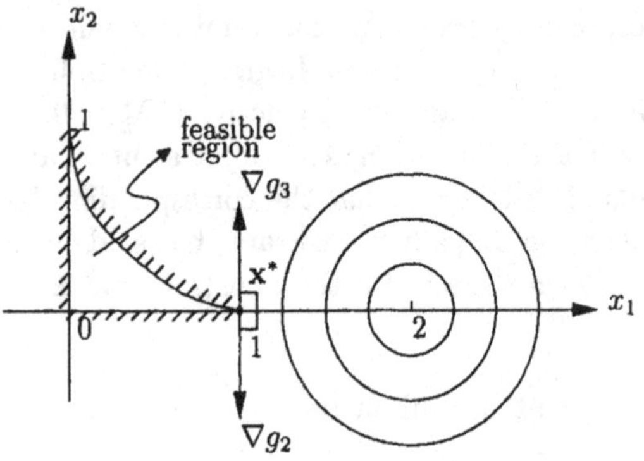

Figure 3.6: Failure of constraint qualification

At this point

$$\nabla g_2 = \begin{bmatrix} 0 \\ -1 \end{bmatrix}, \ \nabla g_3 = \begin{bmatrix} 0 \\ 1 \end{bmatrix}$$

giving $\nabla g_2 = -\nabla g_3$ at $\mathbf{x}^*$ as shown in Figure 3.6.

Thus no $\mathbf{h}$ exists that satisfies the constraint qualification (3.21). Therefore the application of KKT theory to this problem will break down since no $\boldsymbol{\lambda}^*$ exists.

Also note that the Jacobian of the active constraints at $\mathbf{x}^* = [1, 0]^T$:

$$\frac{\partial \mathbf{g}}{\partial \mathbf{x}}(\mathbf{x}^*) = \begin{bmatrix} \dfrac{\partial g_2}{\partial x_1} & \dfrac{\partial g_3}{\partial x_1} \\ \dfrac{\partial g_2}{\partial x_2} & \dfrac{\partial g_3}{\partial x_2} \end{bmatrix} = \begin{bmatrix} 0 & 0 \\ -1 & 1 \end{bmatrix}$$

is of rank $1 < 2$, and therefore not maximal which also indicates that $\boldsymbol{\lambda}^*$ does not exist, and therefore the problem cannot be analysed via the KKT conditions.

3.2.2.4 Simple example of application of KKT conditions

Minimize $f(\mathbf{x}) = (x_1 - 1)^2 + (x_2 - 2)^2$

such that $x_1 \geq 0$, $x_2 \geq 0$, $x_1 + x_2 \leq 2$ and $x_2 - x_1 = 1$.

In standard form the constraints are:

$$-x_1 \leq 0,\ -x_2 \leq 0,\ x_1 + x_2 - 2 \leq 0, \text{ and } x_2 - x_1 - 1 = 0$$

and the Lagrangian, now including the equality constraint with Lagrange multiplier μ as well:

$$L(\mathbf{x}, \boldsymbol{\lambda}, \mu) = (x_1-1)^2+(x_2-2)^2+\lambda_3(x_1+x_2-2)-\lambda_1 x_1-\lambda_2 x_2+\mu(x_2-x_1-1).$$

The KKT conditions are

$$\begin{aligned} 2(x_1 - 1) + \lambda_3 - \lambda_1 - \mu = 0;\ 2(x_2 - 2) + \lambda_3 - \lambda_2 + \mu = 0 \\ -x_1 \leq 0;\ -x_2 \leq 0;\ x_1 + x_2 - 2 \leq 0;\ x_2 - x_1 - 1 = 0 \\ \lambda_3(x_1 + x_2 - 2) = 0;\ \lambda_1 x_1 = 0;\ \lambda_2 x_2 = 0 \\ \lambda_1 \geq 0;\ \lambda_2 \geq 0;\ \lambda_3 \geq 0. \end{aligned}$$

In general the approach to the solution is combinatorial. Try different possibilities and test for contradictions. One possible choice is $\lambda_3 \neq 0$ that implies $x_1 + x_2 - 2 = 0$. This together with the equality constraint gives

$$x_1^* = \tfrac{1}{2},\ x_2^* = \tfrac{3}{2},\ \lambda_3^* = 1,\ \lambda_1^* = \lambda_2^* = 0,\ \mu^* = 0 \text{ and } f(\mathbf{x}^*) = \tfrac{1}{2}.$$

This candidate solution satisfies all the KKT conditions and is indeed the optimum solution. Why? The graphical solution is depicted in Figure 3.7.

3.3 Saddle point theory and duality

3.3.1 Saddle point theorem

If the point $(\mathbf{x}^*, \boldsymbol{\lambda}^*)$, with $\boldsymbol{\lambda}^* \geq \mathbf{0}$, is a saddle point of the Lagrangian associated with the primal problem (3.18) then $\mathbf{x}^*$ is a solution to the primal problem. For a proof of this statement see Theorem 6.4.2 in Chapter 6.

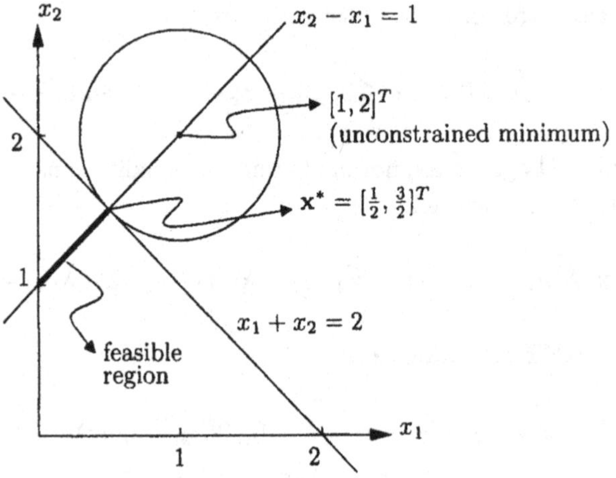

Figure 3.7: Graphical solution to example problem in 3.2.2.4

3.3.2 Duality

Define the *dual function*:

$$h(\boldsymbol{\lambda}) = \min_{\mathbf{x}} L(\mathbf{x}, \boldsymbol{\lambda}). \tag{3.22}$$

Note that the minimizer $\mathbf{x}^*(\boldsymbol{\lambda})$ does not necessarily satisfy $\mathbf{g}(\mathbf{x}) \leq \mathbf{0}$, and indeed the minimum may not even exist for all $\boldsymbol{\lambda}$.

Defining the set

$$D = \{\boldsymbol{\lambda} | h(\boldsymbol{\lambda}) \ \exists \text{ and } \boldsymbol{\lambda} \geq \mathbf{0}\} \tag{3.23}$$

allows for the formulation of the *dual problem* (DP):

$$\underset{\boldsymbol{\lambda} \in D}{\text{maximize}}\, h(\boldsymbol{\lambda}) \tag{3.24}$$

which is equivalent to $\left\{ \max_{\boldsymbol{\lambda} \in D} (\min_{\mathbf{x}} L(\mathbf{x}, \boldsymbol{\lambda})) \right\}$.

3.3.3 Duality theorem

The point $(\mathbf{x}^*, \boldsymbol{\lambda}^*)$, with $\boldsymbol{\lambda}^* \geq \mathbf{0}$, is a saddle point of the Lagrangian function of the primal problem (PP), defined by (3.18), *if and only if*:

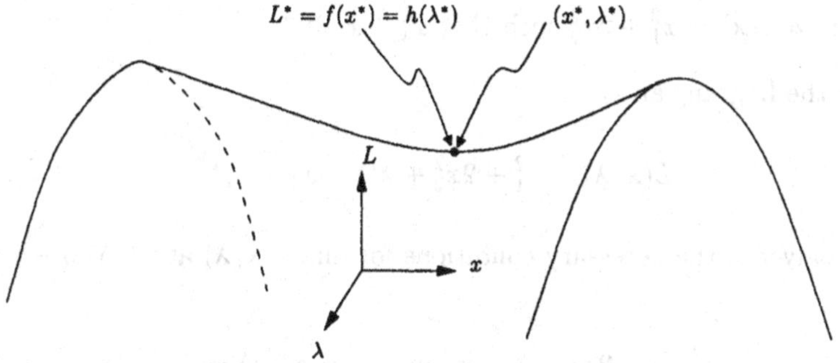

Figure 3.8: Schematic representation of saddle point solution to PP

(i) $\mathbf{x}^*$ is a solution to the primal problem (PP),

(ii) $\boldsymbol{\lambda}^*$ is a solution to the dual problem (DP), and

(iii) $f(\mathbf{x}^*) = h(\boldsymbol{\lambda}^*)$.

A schematic representation of this theorem is given in Figure 3.8 and a formal proof is given in listed Theorem 6.4.4 in Chapter 6.

3.3.3.1 Practical significance of the duality theorem

The implication of the Duality Theorem is that the PP may be solved by carrying out the following steps:

1. If possible, solve the DP separately to give $\boldsymbol{\lambda}^* \geq \mathbf{0}$, i.e. solve an essentially unconstrained problem.
2. With $\boldsymbol{\lambda}^*$ known solve the unconstrained problem: $\min_{\mathbf{x}} L(\mathbf{x}, \boldsymbol{\lambda}^*)$ to give $\mathbf{x}^* = \mathbf{x}^*(\boldsymbol{\lambda}^*)$.
3. Test whether $(\mathbf{x}^*, \boldsymbol{\lambda}^*)$ satisfy the KKT conditions.

3.3.3.2 Example of the application of duality

Consider the problem:

minimize $f(\mathbf{x}) = x_1^2 + 2x_2^2$ such that $x_1 + x_2 \geq 1$.

Here the Lagrangian is

$$L(\mathbf{x}, \lambda) = x_1^2 + 2x_2^2 + \lambda(1 - x_1 - x_2).$$

For a given λ the necessary conditions for $\min_{\mathbf{x}} L(\mathbf{x}, \lambda)$ at $\mathbf{x}^*(\lambda)$ are

$$\begin{aligned} \frac{\partial L}{\partial x_1} &= 2x_1 - \lambda = 0 \quad \Rightarrow \quad x_1 = x_1^*(\lambda) = \tfrac{\lambda}{2} \\ \frac{\partial L}{\partial x_2} &= 4x_2 - \lambda = 0 \quad \Rightarrow \quad x_2 = x_2^*(\lambda) = \tfrac{\lambda}{4}. \end{aligned}$$

Note that $\mathbf{x}^*(\lambda)$ is a minimum since the Hessian of the Lagrangian with respect to $\mathbf{x}$, $\mathbf{H}_L = \begin{bmatrix} 2 & 0 \\ 0 & 4 \end{bmatrix}$ is positive-definite.

Substituting the minimizing values (i.t.o. λ) into L gives

$$L(\mathbf{x}^*(\lambda), \lambda) = h(\lambda) = \left(\tfrac{\lambda}{2}\right)^2 + 2\left(\tfrac{\lambda}{4}\right)^2 + \lambda\left(1 - \tfrac{\lambda}{2} - \tfrac{\lambda}{4}\right)$$

i.e. the dual function is $h(\lambda) = -\frac{3}{8}\lambda^2 + \lambda$.

Now solve the DP: $\max_{\lambda} h(\lambda)$.

The necessary condition for a for maximum is $\frac{dh}{d\lambda} = 0$, i.e.

$$-\tfrac{3}{4}\lambda + 1 = 0 \Rightarrow \lambda^* = \tfrac{4}{3} > 0 \text{ (a maximum since } \tfrac{d^2h}{d\lambda^2} = -\tfrac{3}{4} < 0).$$

Substituting λ^* in $\mathbf{x}^*(\lambda)$ above gives

$$x_1^* = \tfrac{2}{3};\ x_2^* = \tfrac{1}{3} \Rightarrow f(\mathbf{x}^*) = \tfrac{2}{3} = h(\lambda^*).$$

and since $(\mathbf{x}^*, \lambda^*)$, with $\lambda^* = \frac{4}{3} > 0$, clearly satisfies the KKT conditions, it indeed represents the optimum solution.

The dual approach is an important method in some structural optimization problems (Fleury, 1979). It is also employed in the development of the *augmented Lagrange multiplier method* to be discussed later.

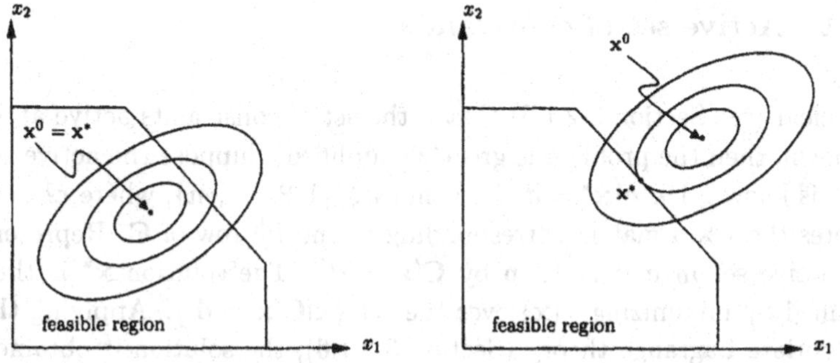

Figure 3.9: The solution of the QP problem: may be an interior point or lie on the boundary of the feasible region

3.4 Quadratic programming

The problem of minimizing a positive-definite quadratic function subject to linear constraints, dealt with in Section 3.2.1.6, is a special case of a *quadratic programming* (QP) problem. Consider now a more general case of a QP problem with inequality constraints:

$$\text{minimize } f(\mathbf{x}) = \tfrac{1}{2}\mathbf{x}^T\mathbf{A}\mathbf{x} + \mathbf{b}^T\mathbf{x} + c$$

subject to

$$\mathbf{C}\mathbf{x} \leq \mathbf{d} \tag{3.25}$$

where $\mathbf{C}$ is a $m \times n$ matrix and $\mathbf{d}$ is a m-vector.

The solution point may be an interior point or may lie on the boundary of the feasible region as shown for the two-dimensional case in Figure 3.9.

If the solution point is an interior point then no constraints are active, and $\mathbf{x}^* = \mathbf{x}^0 = -\mathbf{A}^{-1}\mathbf{b}$ as shown in the figure.

The QP problem is often an important sub-problem to be solved when applying modern methods to more general problems (see the discussion of the SQP method later).

3.4.1 Active set of constraints

It is clear (see Section 3.2.1.6) that if the set of constraints active at $\mathbf{x}^*$ is known, then the problem is greatly simplified. Suppose the active set at $\mathbf{x}^*$ is known, i.e. $\mathbf{c}^j\mathbf{x}^* = d_j$ for some $j \in \{1, 2, \ldots, m\}$, where $\mathbf{c}^j$ here denotes the $1 \times n$ matrix corresponding to the jth row of $\mathbf{C}$. Represent this active set in matrix form by $\mathbf{C}'\mathbf{x} = \mathbf{d}'$. The solution $\mathbf{x}^*$ is then obtained by minimizing $f(\mathbf{x})$ over the set $\{\mathbf{x}|\mathbf{C}'\mathbf{x} = \mathbf{d}'\}$. Applying the appropriate Lagrange theory (Section 3.2.1.6), the solution is obtained by solving the linear system:

$$\begin{bmatrix} \mathbf{A} & \mathbf{C}'^T \\ \mathbf{C}' & \mathbf{0} \end{bmatrix} \begin{bmatrix} \mathbf{x}^* \\ \boldsymbol{\lambda}^* \end{bmatrix} = \begin{bmatrix} -\mathbf{b}^* \\ \mathbf{d}' \end{bmatrix}. \tag{3.26}$$

In solving the QP the major task therefore lies in the *identification of the active set* of constraints.

3.4.2 The method of Theil and Van de Panne

The method of Theil and van de Panne (1961) is a straight-forward method for identifying the active set. A description of the method, after Wismer and Chattergy (1978), for the problem graphically depicted in Figure 3.10 is now given.

Let $V[\mathbf{x}]$ denote the set of constraints violated at $\mathbf{x}$. Select as initial candidate solution the unconstrained minimum $\mathbf{x}^0 = \mathbf{A}^{-1}\mathbf{b}$. Clearly for the example sketched in Figure 3.10, $V[\mathbf{x}^0] = \{1, 2, 3\}$. Therefore $\mathbf{x}^0$ is not the solution. Now consider, as active constraint (set S_1), each constraint in $V[\mathbf{x}^0]$ separately, and let $\mathbf{x}[S_1]$ denote the corresponding minimizer:

$$\begin{aligned}
S_1 = \{1\} \Rightarrow \quad & \mathbf{x}[S_1] = \mathbf{a} \Rightarrow V[\mathbf{a}] = \{2, 3\} \neq \phi \text{ (not empty)} \\
S_1 = \{2\} \Rightarrow \quad & \mathbf{x}[S_1] = \mathbf{b} \Rightarrow V[\mathbf{b}] = \{3\} \neq \phi \\
S_1 = \{3\} \Rightarrow \quad & \mathbf{x}[S_1] = \mathbf{c} \Rightarrow V[\mathbf{c}] = \{1, 2\} \neq \phi.
\end{aligned}$$

Since all the solutions with a single active constraint violate one or more constraints, the next step is to consider different combinations S_2 of two

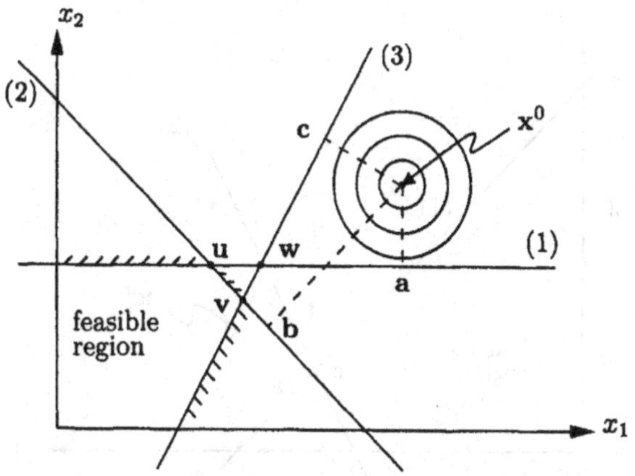

Figure 3.10: Graphical illustration of the method of Theil and Van de Panne

simultaneously active constraints from $V[\mathbf{x}^0]$:

$$\begin{aligned} S_2 = \{1,2\} \Rightarrow &\quad \mathbf{x}[S_2] = \mathbf{u} \Rightarrow V[\mathbf{u}] = \phi \\ S_2 = \{1,3\} \Rightarrow &\quad \mathbf{x}[S_2] = \mathbf{w} \Rightarrow V[\mathbf{w}] = \{2\} \\ S_2 = \{2,3\} \Rightarrow &\quad \mathbf{x}[S_2] = \mathbf{v} \Rightarrow V[\mathbf{v}] = \phi. \end{aligned}$$

Since both $V[\mathbf{u}]$ and $V[\mathbf{v}]$ are empty, $\mathbf{u}$ and $\mathbf{v}$ are both candidate solutions. Apply the KKT conditions to both $\mathbf{u}$ and $\mathbf{v}$ separately, to determine which point is the optimum one. Assume it can be shown, from the solution of (3.26) that for $\mathbf{u}$: $\lambda_1 < 0$ (which indeed is apparently so from the geometry of the problem sketched in Figure 3.10), then it follows that $\mathbf{u}$ is non-optimal. On the other hand, assume it can be shown that for $\mathbf{v}$: $\lambda_2 > 0$ and $\lambda_3 > 0$ (which is evidently the case in the figure), then $\mathbf{v}$ is optimum, i.e. $\mathbf{x}^* = \mathbf{v}$.

3.4.2.1 Explicit example

Solve by means of the method of *Theil and van de Panne* the following QP problem:

$$\text{minimize}\, f(\mathbf{x}) = \tfrac{1}{2}x_1^2 - x_1x_2 + x_2^2 - 2x_1 + x_2$$

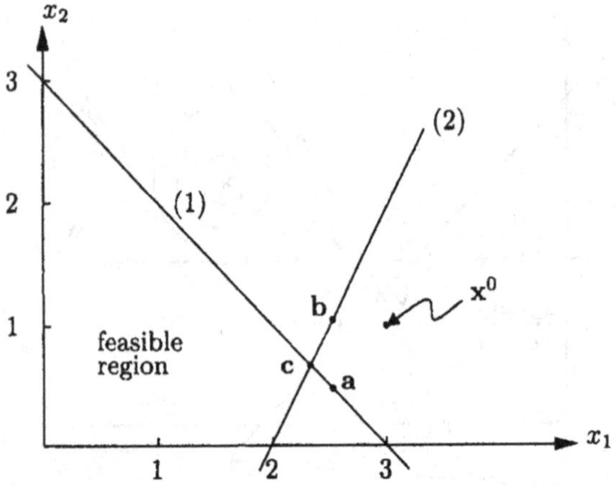

Figure 3.11: Graphical solution to explicit example 3.4.2.1

such that

$$x_1 \geq 0;\ x_2 \geq 0;\ x_1 + x_2 \leq 3;\ 2x_1 - x_2 \leq 4.$$

In matrix form $f(\mathbf{x})$ is given by $f(\mathbf{x}) = \frac{1}{2}\mathbf{x}^T\mathbf{A}\mathbf{x} + \mathbf{b}^T\mathbf{x}$, with $\mathbf{A} = \begin{bmatrix} 1 & -1 \\ -1 & 2 \end{bmatrix}$ and $\mathbf{b} = [-2, 1]^T$.

The unconstrained solution is $\mathbf{x}^0 = \mathbf{A}^{-1}\mathbf{b} = [3, 1]^T$. Clearly $V[\mathbf{x}^0] = \{1, 2\}$ and therefore $\mathbf{x}^0$ is not the solution. Continuing, the method yields:

$$\begin{aligned} S_1 = \{1\} \Rightarrow \quad & \mathbf{x}[S_1] = \mathbf{a} \Rightarrow V[\mathbf{a}] = \{2\} \neq \phi \\ S_1 = \{2\} \Rightarrow \quad & \mathbf{x}[S_1] = \mathbf{b} \Rightarrow V[\mathbf{b}] = \{1\} \neq \phi \\ S_2 = \{1, 2\} \Rightarrow \quad & \mathbf{x}[S_2] = \mathbf{c} \Rightarrow V[\mathbf{c}] = \phi, \text{ where } \mathbf{c} = \left[\tfrac{7}{3}, \tfrac{2}{3}\right]^T. \end{aligned}$$

Applying the KKT conditions to $\mathbf{c}$ establishes the optimality of $\mathbf{c}$ since $\lambda_1 = \lambda_2 = \frac{1}{9} > 0$.

3.5 Modern methods for constrained optimization

The most established gradient-based methods for constrained optimization are

(i) gradient projection methods (Rosen, 1960, 1961),

(ii) augmented Lagrangian multiplier methods (see Haftka and Gündel, 1992), and

(iii) successive or sequential quadratic programming (SQP) methods (see Bazaraa et al., 1993).

All these methods are largely based on the theory already presented here. SQP methods are currently considered to represent the state-of-the-art gradient-based approach to the solution of constrained optimization problems.

3.5.1 The gradient projection method

3.5.1.1 Equality constrained problems

The gradient projection method is due to Rosen (1960, 1961). Consider the linear equality constrained problem:

$$\begin{aligned} \text{minimize} \quad & f(\mathbf{x}) \\ \text{such that} \quad & \mathbf{A}\mathbf{x} - \mathbf{b} = \mathbf{0} \end{aligned} \tag{3.27}$$

where $\mathbf{A}$ is a $r \times n$ matrix, $r < n$ and $\mathbf{b}$ a r-vector.

The gradient projection method for solving (3.27) is based on the following argument.

Assume that $\mathbf{x}'$ is feasible, i.e. $\mathbf{A}\mathbf{x}' - \mathbf{b} = \mathbf{0}$ in Figure 3.12. A direction $\mathbf{s}$, ($\|\mathbf{s}\| = 1$) is sought such that a step $\alpha\mathbf{s}$ ($\alpha > 0$) from $\mathbf{x}'$ in the direction $\mathbf{s}$ also gives a feasible point, i.e. $\mathbf{A}(\mathbf{x}' + \alpha\mathbf{s}) - \mathbf{b} = \mathbf{0}$. This condition reduces to

$$\mathbf{A}\mathbf{s} = \mathbf{0}. \tag{3.28}$$

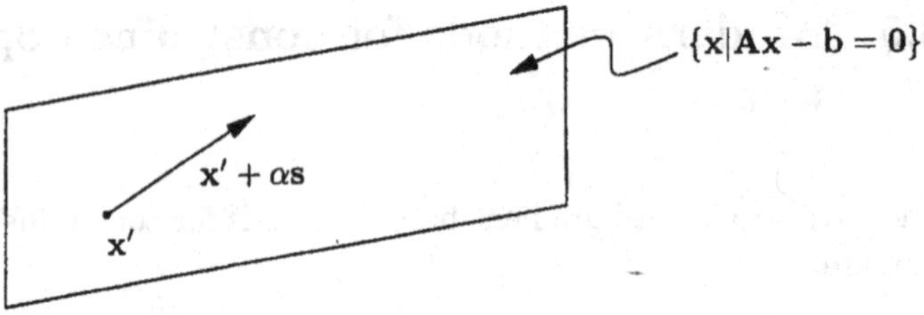

Figure 3.12: Schematic representation of the subspace $\{\mathbf{x}|\mathbf{Ax} - \mathbf{b} = \mathbf{0}\}$

Also since $\|\mathbf{s}\| = 1$, it also follows that

$$1 - \mathbf{s}^T\mathbf{s} = 0. \tag{3.29}$$

It is now required that $\mathbf{s}$ be chosen such that it corresponds to the direction which gives the steepest descent at $\mathbf{x}'$ subject to satisfying the constraints specified by (3.28) and (3.29). This requirement is equivalent to determining a $\mathbf{s}$ such that the directional derivative at $\mathbf{x}'$:

$$R(\mathbf{s}) = \left.\frac{dF(0)}{d\alpha}\right|_{\mathbf{s}} = \boldsymbol{\nabla}^T f(\mathbf{x}')\mathbf{s} \tag{3.30}$$

is minimized with respect to $\mathbf{s}$, where $F(\alpha) = f(\mathbf{x}' + \alpha\mathbf{s})$.

Applying the classical Lagrange theory for minimizing a function subject to equality constraints, requires the formulation of the following Lagrangian function:

$$L(\mathbf{s}, \boldsymbol{\lambda}, \lambda_0) = \boldsymbol{\nabla}^T f(\mathbf{x}')\mathbf{s} + \boldsymbol{\lambda}^T\mathbf{As} + \lambda_0(1 - \mathbf{s}^T\mathbf{s}) \tag{3.31}$$

where the variables $\mathbf{s} = [s_1, s_2, \ldots, s_n]^T$ correspond to the direction cosines of $\mathbf{s}$. The Lagrangian necessary conditions for the constrained minimum are

$$\begin{aligned}
\boldsymbol{\nabla}_{\mathbf{s}} L &= \boldsymbol{\nabla} f(\mathbf{x}') + \mathbf{A}^T\boldsymbol{\lambda} - 2\lambda_0\mathbf{s} = \mathbf{0} && (3.32)\\
\boldsymbol{\nabla}_{\boldsymbol{\lambda}} L &= \mathbf{As} = \mathbf{0} && (3.33)\\
\boldsymbol{\nabla}_{\lambda_0} L &= (1 - \mathbf{s}^T\mathbf{s}) = 0. && (3.34)
\end{aligned}$$

Equation (3.32) yields

$$\mathbf{s} = \frac{1}{2\lambda_0}(\boldsymbol{\nabla} f(\mathbf{x}') + \mathbf{A}^T\boldsymbol{\lambda}). \tag{3.35}$$

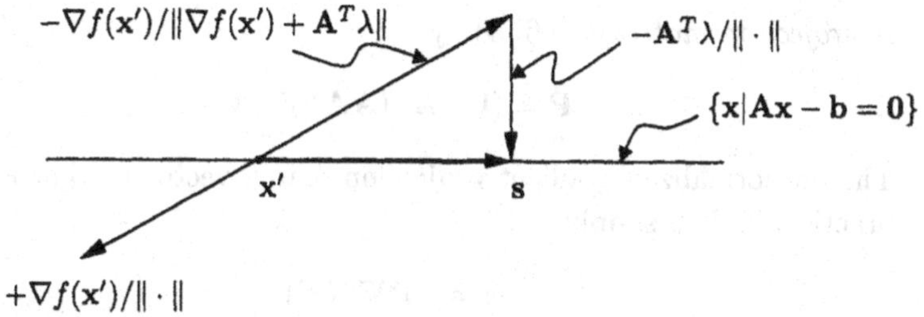

Figure 3.13: Schematic representation of projected direction of steepest descent $\mathbf{s}$

Substituting (3.35) into (3.34) gives

$$1 = \frac{1}{4\lambda_0^2}(\nabla f(\mathbf{x}') + \mathbf{A}^T\boldsymbol{\lambda})^T(\nabla f(\mathbf{x}') + \mathbf{A}^T\boldsymbol{\lambda})$$

and thus

$$\lambda_0 = \pm\frac{1}{2}\|\nabla f(\mathbf{x}') + \mathbf{A}^T\boldsymbol{\lambda}\|. \tag{3.36}$$

Substituting (3.36) into (3.35) gives $\mathbf{s} = \pm(\nabla f(\mathbf{x}') + \mathbf{A}^T\boldsymbol{\lambda})/\|\nabla f(\mathbf{x}') + \mathbf{A}^T\boldsymbol{\lambda}\|$.

For maximum descent choose the negative sign as shown in Figure 3.13. This also ensures that the Hessian of the Lagrangian with respect to $\mathbf{s}$ is positive-definite (sufficiency condition) which ensures that $R(\mathbf{s})$ in (3.30) indeed assumes a minimum value with respect to $\mathbf{s}$. Thus the constrained (projected) direction of steepest descent is chosen as

$$\mathbf{s} = -(\nabla f(\mathbf{x}') + \mathbf{A}^T\boldsymbol{\lambda})/\|\nabla f(\mathbf{x}') + \mathbf{A}^T\boldsymbol{\lambda}\|. \tag{3.37}$$

It remains to solve for $\boldsymbol{\lambda}$. Equations (3.33) and (3.37) imply $\mathbf{A}(\nabla f(\mathbf{x}') + \mathbf{A}^T\boldsymbol{\lambda}) = \mathbf{0}$. Thus if $\mathbf{s} \neq \mathbf{0}$ then $\mathbf{A}\mathbf{A}^T\boldsymbol{\lambda} = -\mathbf{A}\nabla f(\mathbf{x}')$ with solution

$$\boldsymbol{\lambda} = -(\mathbf{A}\mathbf{A}^T)^{-1}\mathbf{A}\nabla f(\mathbf{x}'). \tag{3.38}$$

The direction $\mathbf{s}$, called the *gradient projection* direction, is therefore finally given by

$$\mathbf{s} = -(\mathbf{I} - \mathbf{A}^T(\mathbf{A}\mathbf{A}^T)^{-1}\mathbf{A})\nabla f(\mathbf{x}')/\|\nabla f(\mathbf{x}') + \mathbf{A}^T\boldsymbol{\lambda}\|. \tag{3.39}$$

A *projection matrix* is defined by

$$\mathbf{P} = (\mathbf{I} - \mathbf{A}^T(\mathbf{A}\mathbf{A}^T)^{-1}\mathbf{A}). \tag{3.40}$$

The un-normalized gradient projection search vector $\mathbf{u}$, that is used in practice, is then simply

$$\mathbf{u} = -\mathbf{P}\nabla f(\mathbf{x}'). \tag{3.41}$$

3.5.1.2 Extension to non-linear constraints

Consider the more general problem:

$$\begin{aligned} &\text{minimize} && f(\mathbf{x}) \\ &\text{such that} && h_i(\mathbf{x}) = 0,\ i = 1, 2, \dots, r, \end{aligned} \tag{3.42}$$

or in vector form $\mathbf{h}(\mathbf{x}) = \mathbf{0}$, where the constraints may be non-linear.

Linearize the constraint functions $h_i(\mathbf{x})$ at the feasible point $\mathbf{x}'$, by the truncated Taylor expansions:

$$h_i(\mathbf{x}) = h_i(\mathbf{x}' + (\mathbf{x} - \mathbf{x}')) \cong h_i(\mathbf{x}') + \nabla^T h_i(\mathbf{x}')(\mathbf{x} - \mathbf{x}')$$

which allows for the following approximations to the constraints:

$$\nabla^T h_i(\mathbf{x}')(\mathbf{x} - \mathbf{x}') = 0,\ i = 1, 2, \dots, r \tag{3.43}$$

in the neighbourhood of $\mathbf{x}'$, since the $h_i(\mathbf{x}') = 0$. This set of linearized constraints may be written in matrix form as

$$\left[\frac{\partial \mathbf{h}(\mathbf{x}')}{\partial \mathbf{x}}\right]^T \mathbf{x} - \mathbf{b} = \mathbf{0}$$

where

$$\mathbf{b} = \left[\frac{\partial \mathbf{h}(\mathbf{x}')}{\partial \mathbf{x}}\right]^T \mathbf{x}'. \tag{3.44}$$

The *linearized problem* at $\mathbf{x}'$ therefore becomes

$$\text{minimize } f(\mathbf{x}) \text{ such that } \mathbf{A}\mathbf{x} - \mathbf{b} = \mathbf{0} \tag{3.45}$$

where $\mathbf{A} = \left[\frac{\partial \mathbf{h}(\mathbf{x}')}{\partial \mathbf{x}}\right]^T$ and $\mathbf{b} = \left[\frac{\partial \mathbf{h}(\mathbf{x}')}{\partial \mathbf{x}}\right]^T \mathbf{x}'$.

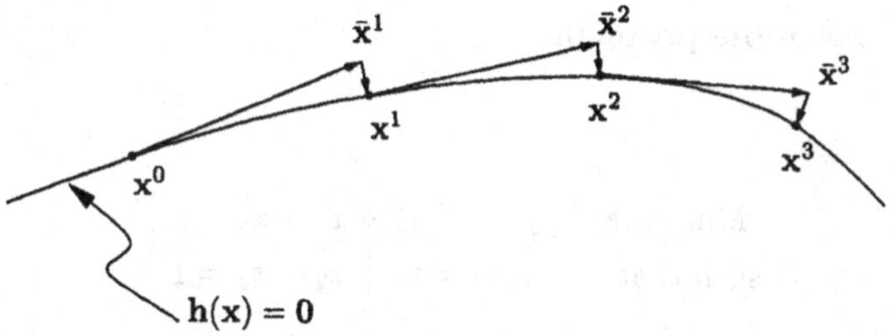

Figure 3.14: Schematic representation of correction steps when applying the gradient projection method to nonlinear constrained problems

Since the problem is now linearized, the computation of the gradient projection direction at $\mathbf{x}'$ is identical to that before, i.e.:

$\mathbf{u} = \mathbf{P}(\mathbf{x}')\boldsymbol{\nabla} f(\mathbf{x}')$, but the projection matrix $\mathbf{P}(\mathbf{x}') = (\mathbf{I} - \mathbf{A}^T(\mathbf{A}\mathbf{A}^T)^{-1}\mathbf{A})$ is now dependent on $\mathbf{x}'$, since $\mathbf{A}$ is given by $\left[\dfrac{\partial \mathbf{h}(\mathbf{x}')}{\partial \mathbf{x}}\right]^T$.

For an initial feasible point $\mathbf{x}^0 (= \mathbf{x}')$ a new point in the gradient projection direction of descent $\mathbf{u}^1 = -\mathbf{P}(\mathbf{x}^0)\boldsymbol{\nabla} f(\mathbf{x}^0)$, is $\bar{\mathbf{x}}^1 = \mathbf{x}^0 + \alpha_1 \mathbf{u}^1$ for step size $\alpha_1 > 0$ as shown in Figure 3.14.

In general $\mathbf{h}(\bar{\mathbf{x}}^1) \neq 0$, and a correction step must be calculated: $\bar{\mathbf{x}}^1 \to \mathbf{x}^1$. How is this correction step computed? Clearly (see Figure 3.14), the step should be such that its projection at $\mathbf{x}^1$ is zero, i.e. $\mathbf{P}(\mathbf{x}^1)(\mathbf{x}^1 - \bar{\mathbf{x}}^1) = \mathbf{0}$ and also $\mathbf{h}(\mathbf{x}^1) = \mathbf{0}$. These two conditions imply that

$$(\mathbf{I} - \mathbf{A}^T(\mathbf{A}\mathbf{A}^T)^{-1}\mathbf{A})(\mathbf{x}^1 - \bar{\mathbf{x}}^1) = \mathbf{0}$$

with $\mathbf{A}$ evaluated at $\mathbf{x}^1$, which gives

$$\mathbf{x}^1 \cong \bar{\mathbf{x}}^1 - \mathbf{A}^T(\mathbf{A}\mathbf{A}^T)^{-1}\mathbf{h}(\bar{\mathbf{x}}^1) \tag{3.46}$$

as the correction step, where use was made of the expression $\mathbf{h}(\mathbf{x}) \cong \mathbf{A}\mathbf{x} - \mathbf{b}$ for both $\mathbf{h}(\bar{\mathbf{x}}^1)$ and $\mathbf{h}(\mathbf{x}^1)$ and of the fact that $\mathbf{h}(\mathbf{x}^1) = \mathbf{0}$.

Since the correction step is based on an approximation it may have to be applied repeatedly until $\mathbf{h}(\mathbf{x}^1)$ is sufficiently small. Having found a satisfactory $\mathbf{x}^1$, the procedure is repeated successively for $k = 2, 3, \ldots$ to give $\mathbf{x}^2, \mathbf{x}^3, \ldots$, until $\mathbf{P}(\mathbf{x}^k)\boldsymbol{\nabla} f(\mathbf{x}^k) \cong \mathbf{0}$.

3.5.1.3 Example problem

$$\begin{array}{ll} \text{Minimize} & f(\mathbf{x}) = x_1^2 + x_2^2 + x_3^2 \\ \text{such that} & h(\mathbf{x}) = x_1 + x_2 + x_3 = 1 \end{array}$$

with initial feasible point $\mathbf{x}^0 = [1, 0, 0]^T$.

First evaluate the projection matrix $\mathbf{P} = \mathbf{I} - \mathbf{A}^T(\mathbf{A}\mathbf{A}^T)^{-1}\mathbf{A}$.

Here $\mathbf{A} = [1\ 1\ 1]$, $\mathbf{A}\mathbf{A}^T = 3$ and $(\mathbf{A}\mathbf{A}^T)^{-1} = \frac{1}{3}$, thus giving

$$\mathbf{P} = \begin{bmatrix} 1 & 0 & 0 \\ 0 & 1 & 0 \\ 0 & 0 & 1 \end{bmatrix} - \frac{1}{3}\begin{bmatrix} 1 & 1 & 1 \\ 1 & 1 & 1 \\ 1 & 1 & 1 \end{bmatrix} = \frac{1}{3}\begin{bmatrix} 2 & -1 & -1 \\ -1 & 2 & -1 \\ -1 & -1 & 2 \end{bmatrix}$$

The gradient vector is $\nabla f(\mathbf{x}) = \begin{bmatrix} 2x_1 \\ 2x_2 \\ 2x_3 \end{bmatrix}$ giving at $\mathbf{x}^0$, $\nabla f(\mathbf{x}) = \begin{bmatrix} 2 \\ 0 \\ 0 \end{bmatrix}$.

The search at direction at $\mathbf{x}^0$ is therefore given by

$$-\mathbf{P}\nabla f(\mathbf{x}^0) = -\frac{1}{3}\begin{bmatrix} 2 & -1 & -1 \\ -1 & 2 & -1 \\ -1 & -1 & 2 \end{bmatrix}\begin{bmatrix} 2 \\ 0 \\ 0 \end{bmatrix} = -\frac{1}{3}\begin{bmatrix} 4 \\ -2 \\ -2 \end{bmatrix} = \frac{2}{3}\begin{bmatrix} -2 \\ 1 \\ 1 \end{bmatrix}$$

or more conveniently, for this example, choose thc search direction simply as $\mathbf{u} = \begin{bmatrix} -2 \\ 1 \\ 1 \end{bmatrix}$.

For a suitable value of λ the next point is given by

$$\mathbf{x}^1 = \mathbf{x}^0 + \lambda\begin{bmatrix} -2 \\ 1 \\ 1 \end{bmatrix} = \begin{bmatrix} 1 - 2\lambda \\ \lambda \\ \lambda \end{bmatrix}.$$

Substituting the above in $f(\mathbf{x}^1) = f(\mathbf{x}^0 + \lambda\mathbf{u}) = F(\lambda) = (1 - 2\lambda)^2 + \lambda^2 + \lambda^2$, it follows that for optimal descent in the direction $\mathbf{u}$:

$$\frac{dF}{d\lambda} = -2(1 - 2\lambda)2 + 2\lambda + 2\lambda = 0 \Rightarrow \lambda = \frac{1}{3}$$

which gives

$$\mathbf{x}^1 = \left[\tfrac{1}{3}, \tfrac{1}{3}, \tfrac{1}{3}\right]^T, \ \nabla f_1(\mathbf{x}^1) = \left[\tfrac{2}{3}, \tfrac{2}{3}, \tfrac{2}{3}\right]^T$$

and

$$\mathbf{P}\nabla f(\mathbf{x}^1) = \tfrac{1}{3}\begin{bmatrix} 2 & -1 & -1 \\ -1 & 2 & -1 \\ -1 & -1 & 2 \end{bmatrix} \tfrac{2}{3}\begin{bmatrix} 1 \\ 1 \\ 1 \end{bmatrix} = \tfrac{2}{9}\begin{bmatrix} 0 \\ 0 \\ 0 \end{bmatrix} = \mathbf{0}.$$

Since the projection of the gradient vector at $\mathbf{x}^1$ is zero it is the optimum point.

3.5.1.4 Extension to linear inequality constraints

Consider the case of linear inequality constraints:

$$\mathbf{A}\mathbf{x} - \mathbf{b} \leq \mathbf{0} \tag{3.47}$$

where $\mathbf{A}$ is a $m \times n$ matrix and $\mathbf{b}$ a m-vector, i.e. the individual constraints are of the form

$$g_j(\mathbf{x}) = \boldsymbol{a}^j\mathbf{x} - b_j \leq 0, \ j = 1, 2, \ldots, m.$$

where $\boldsymbol{a}^j$ denotes the $1 \times n$ matrix corresponding to the jth row of $\mathbf{A}$. Suppose at the current point $\mathbf{x}^k$, $r(\leq m)$ constraints are enforced, i.e. are active. Then a set of equality constraints, corresponding to the active constraints in (3.47) apply at $\mathbf{x}^k$, i.e. $\mathbf{A}_r\mathbf{x}^k - \mathbf{b}^r = \mathbf{0}$, where $\mathbf{A}_r$ and $\mathbf{b}^r$ correspond to the set of r active constraints in (3.47).

Now apply the gradient projection method as depicted in Figure 3.15 where the recursion is as follows:

$$\mathbf{u}^{k+1} = -\mathbf{P}(\mathbf{x}^k)\nabla f(\mathbf{x}^k) \text{ and } \mathbf{x}^{k+1} = \mathbf{x}^k + \alpha_{k+1}\mathbf{u}^{k+1}$$

where

$$f(\mathbf{x}^k + \alpha_{k+1}\mathbf{u}^{k+1}) = \min_{\alpha} f(\mathbf{x}^k + \alpha\mathbf{u}^{k+1}). \tag{3.48}$$

Two possibilities may arise:

(i) No additional constraint is encountered along $\mathbf{u}^{k+1}$ before $\mathbf{x}^{k+1} = \mathbf{x}^k + \alpha_{k+1}\mathbf{u}^{k+1}$. Test whether $\mathbf{P}\nabla f(\mathbf{x}^{k+1}) = \mathbf{0}$. If so then $\mathbf{x}^* = \mathbf{x}^{k+1}$, otherwise set $\mathbf{u}^{k+2} = -\mathbf{P}(\mathbf{x}^{k+1})\nabla f(\mathbf{x}^{k+1})$ and continue.

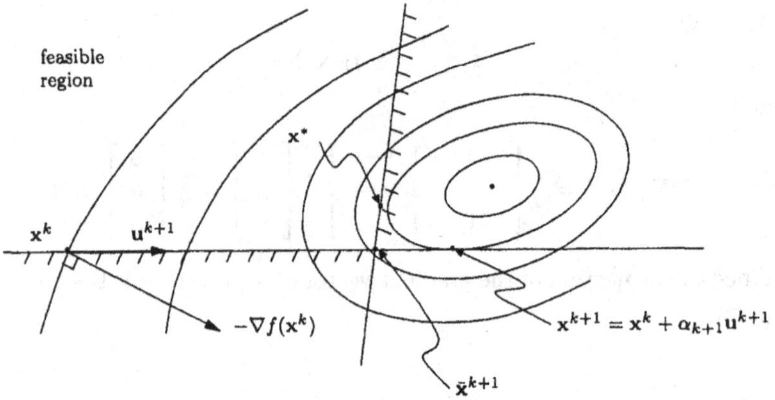

Figure 3.15: Representation of the gradient projection method for linear inequality constraints

(ii) If an additional constraint is encountered before $\mathbf{x}^{k+1} = \mathbf{x}^k + \alpha_{k+1}\mathbf{u}^{k+1}$ at $\overline{\mathbf{x}}^{k+1}$, (see Figure 3.15), then set $\mathbf{x}^{k+1} = \overline{\mathbf{x}}^{k+1}$ and add new constraint to active set, with associated matrix $\mathbf{A}_{r+1}$. Compute new $\mathbf{P}$ and set $\mathbf{u}^{k+2} = -\mathbf{P}(\mathbf{x}^{k+1})\nabla f(\mathbf{x}^{k+1})$. Continue this process until for some active set at $\mathbf{x}^p$, $\mathbf{P}\nabla f(\mathbf{x}^p) = \mathbf{0}$.

How do we know that if $\mathbf{P}\nabla f(\mathbf{x}^k) = \mathbf{0}$ occurs, that all the identified constraints are active? The answer is given by the following argument.

If $\mathbf{P}\nabla f(\mathbf{x}^k) = \mathbf{0}$ it implies that $(\mathbf{I} - \mathbf{A}^T(\mathbf{A}\mathbf{A}^T)^{-1}\mathbf{A})\nabla f(\mathbf{x}^k) = \mathbf{0}$ which, using expression (3.38), is equivalent to $\nabla f(\mathbf{x}^k) + \mathbf{A}^T\boldsymbol{\lambda} = \mathbf{0}$, i.e. $\nabla f(\mathbf{x}^k) + \sum \lambda_i \nabla g_i(\mathbf{x}^k) = \mathbf{0}$ for all $i \in I_a$ = set of all active constraints. This expression is nothing else but the KKT conditions for the optimum at $\mathbf{x}^k$, provided that $\lambda_i \geq 0$ for all $i \in I_a$. Now, if $\mathbf{P}\nabla f(\mathbf{x}^k) = \mathbf{0}$ occurs, then if $\lambda_i < 0$ for some i, remove the corresponding constraint from the active set. In practice remove the constraint with the most negative multiplier, compute the new projection matrix $\mathbf{P}$, and continue.

3.5.2 Multiplier methods

These methods combine the classical Lagrangian method with the penalty function approach. In the Lagrangian approach the minimum point of the constrained problem coincides with a stationary point $(\mathbf{x}^*, \boldsymbol{\lambda}^*)$ of the Lagrangian function which, in general, is *difficult to determine analytically*. On the other hand in the penalty function approach the constrained minimum approximately coincides with the minimum of the penalty function. If, however, high accuracy is required the problem *becomes ill-conditioned.*

In the *multiplier methods* (see Bertsekas, 1976) both the above approaches are combined to give an unconstrained problem which is not ill-conditioned.

As an introduction to the multiplier method consider the equality constrained problem:

$$\begin{aligned} &\text{minimize} && f(\mathbf{x}) \\ &\text{such that} && h_j(\mathbf{x}) = 0,\ j = 1, 2, \ldots, r. \end{aligned} \tag{3.49}$$

The *augmented Lagrange function* $\mathcal{L}$ is introduced as

$$\mathcal{L}(\mathbf{x}, \boldsymbol{\lambda}, \rho) = f(\mathbf{x}) + \sum_{j=1}^{r} \lambda_j h_j(\mathbf{x}) + \rho \sum_{j=1}^{r} h_j^2(\mathbf{x}). \tag{3.50}$$

If all the multipliers λ_j are chosen equal to zero, $\mathcal{L}$ becomes the usual external penalty function. On the other hand, if all the stationary values λ_j^* are available, then *it can be shown* (Fletcher, 1987) that for any positive value of ρ, the minimization of $\mathcal{L}(\mathbf{x}, \boldsymbol{\lambda}^*, \rho)$ with respect to $\mathbf{x}$ gives the solution $\mathbf{x}^*$ to problem (3.49). This result is not surprising since it can be shown that the classical Lagrangian function $L(\mathbf{x}, \boldsymbol{\lambda})$ has a saddle point at $(\mathbf{x}^*, \boldsymbol{\lambda}^*)$.

The multiplier methods are based on the use of approximations to the Lagrange multipliers. If $\boldsymbol{\lambda}^k$ is a good approximation to $\boldsymbol{\lambda}^*$ then it is possible to approach the optimum through the unconstrained minimization of $\mathcal{L}(\mathbf{x}, \boldsymbol{\lambda}^k, \rho)$ without using large values of ρ. The value of ρ must only be sufficiently large to ensure that $\mathcal{L}$ has a local minimum point with respect to $\mathbf{x}$ rather than simply a stationary point at the optimum.

How is the approximation to the Lagrange multiplier vector $\boldsymbol{\lambda}^k$ obtained? To answer this question, compare the stationary conditions with respect to $\mathbf{x}$ for $\mathcal{L}$ (the augmented Lagrangian) with those for L (the classical Lagrangian) at $\mathbf{x}^*$.

For $\mathcal{L}$:

$$\frac{\partial \mathcal{L}}{\partial x_i} = \frac{\partial f}{\partial x_i} + \sum_{j=1}^{r}(\lambda_j^k + 2\rho h_j)\frac{\partial h_j}{\partial x_j} = 0, \; i = 1, 2, \ldots, n. \tag{3.51}$$

For L:

$$\frac{\partial L}{\partial x_i} = \frac{\partial f}{\partial x_i} + \sum_{j=1}^{r}\lambda_j^*\frac{\partial h_j}{\partial x_j} = 0, \; i = 1, 2, \ldots, n. \tag{3.52}$$

The comparison clearly indicates that as the minimum point of $\mathcal{L}$ tends to $\mathbf{x}^*$ that

$$\lambda_j^k + 2\rho h_j \to \lambda_j^*. \tag{3.53}$$

This observation prompted Hestenes (1969) to suggest the following scheme for approximating $\boldsymbol{\lambda}^*$. For a given approximation $\boldsymbol{\lambda}^k$, $k = 1, 2, \ldots,$ minimize $\mathcal{L}(\mathbf{x}, \boldsymbol{\lambda}^k, \rho)$ by some standard unconstrained minimization technique to give a minimum $\mathbf{x}^{*k}$. The components of the new estimate to $\boldsymbol{\lambda}^*$, as suggested by (3.53), is then given by

$$\lambda_j^* \cong \lambda_j^{k+1} = \lambda_j^k + 2\rho h_j(\mathbf{x}^{*k}). \tag{3.54}$$

The value of the penalty parameter $\rho (= \rho_k)$ may also be adjusted from iteration to iteration.

3.5.2.1 Example

Minimize $f(\mathbf{x}) = x_1^2 + 10x_2^2$ such that $h(\mathbf{x}) = x_1 + x_2 - 4 = 0$.

Here $\mathcal{L} = x_1^2 + 10x_2^2 + \lambda(x_1 + x_2 - 4) + \rho(x_1 + x_2 - 4)^2$.

The first order necessary conditions for a constrained minimum with respect to $\mathbf{x}$ for any given λ are

$$\begin{aligned} 2x_1 + \lambda + 2\rho(x_1 + x_2 - 4) &= 0 \\ 20x_2 + \lambda + 2\rho(x_1 + x_2 - 4) &= 0 \end{aligned}$$

from which it follows that

$$x_1 = 10x_2 = \frac{-5\lambda + 40\rho}{10 + 11\rho}.$$

Taking $\lambda^1 = 0$ and $\rho_1 = 1$ gives $\mathbf{x}^{*1} = [1.905, 0.1905]^T$ and $h(\mathbf{x}^{*1}) = -1.905$.

Using the approximation scheme (3.54) for λ^* gives $\lambda^2 = 0{+}2(1)(-1.905) = -3.81$.

Now repeat the minimization of $\mathcal{L}$ with $\lambda^2 = -3.81$ and $\rho_2 = 10$. This gives $\mathbf{x}^{*2} = [3.492, 0.3492]^T$ and $h(\mathbf{x}^{*2}) = -0.1587$, resulting in the new approximation: $\lambda^3 = -3.81 + 2(10)(-0.1587) = -6.984$.

Using λ^3 in the next iteration with $\rho_3 = 10$ gives $\mathbf{x}^{*3} = [3.624, 0.3624]^T$, $h(\mathbf{x}^{*3}) = 0.0136$, which shows good convergence to the exact solution $[3.6363, 0.3636]^T$ without the need for increasing ρ.

3.5.2.2 The multiplier method extended to inequality constraints

The multiplier method can be extended to apply to inequality constraints (Fletcher, 1975).

Consider the problem:

$$\begin{aligned} &\text{minimize} && f(\mathbf{x}) \\ &\text{such that} && g_j(\mathbf{x}) \leq 0,\ j = 1, 2, \ldots, m. \end{aligned} \tag{3.55}$$

Here the augmented Lagrangian function is

$$\mathcal{L}(\mathbf{x}, \boldsymbol{\lambda}, \rho) = f(\mathbf{x}) + \rho \sum_{j=1}^{m} \left\langle \frac{\lambda_j}{2\rho} + g_j(\mathbf{x}) \right\rangle^2 \tag{3.56}$$

where $\langle a \rangle = \max(a, 0)$.

In this case the stationary conditions at $\mathbf{x}^*$ for the augmented and classical Lagrangians are respectively

$$\frac{\partial \mathcal{L}}{\partial x_i} = \frac{\partial f}{\partial x_i} + 2\rho \sum_{j=1}^{m} \left\langle \frac{\lambda_j}{2\rho} + g_j \right\rangle \frac{\partial g_j}{\partial x_i} = 0,\ i = 1, 2, \ldots, n \tag{3.57}$$

and

$$\frac{\partial L}{\partial x_i} = \frac{\partial f}{\partial x_i} + \sum_{j=1}^{m} \lambda_j^* \frac{\partial g_j}{\partial x_i} = 0, \; i = 1, 2, \ldots, n. \tag{3.58}$$

The latter classical KKT conditions require in addition that $\lambda_j^* g_j(\mathbf{x}^*) = 0, j = 1, 2, \ldots, m$.

A comparison of conditions (3.57) and (3.58) leads to the following iterative approximation scheme for $\boldsymbol{\lambda}^*$:

$$\lambda_j^* \cong \lambda_j^{k+1} = \langle \lambda_j^k + 2\rho_k g_j(\mathbf{x}^{*k}) \rangle \tag{3.59}$$

where $\mathbf{x}^{*k}$ minimizes $\mathcal{L}(\mathbf{x}, \boldsymbol{\lambda}^k, \rho_k)$.

3.5.2.3 Example problem with inequality constraint

Minimize $f(\mathbf{x}) = x_1^2 + 10x_2^2$ such that $g(\mathbf{x}) = 4 - x_1 - x_2 \leq 0$

Here $\mathcal{L}(\mathbf{x}, \lambda, \rho) = x_1^2 + 10x_2^2 + \rho\left\langle \frac{\lambda}{2\rho} + (4 - x_1 - x_2)\right\rangle^2$

Now perform the *unconstrained* minimization of $\mathcal{L}$ with respect to $\mathbf{x}$ for given λ and ρ. This is usually done by some standard method such as Fletcher-Reeves or BFGS but here, because of the simplicity of the illustrative example, the necessary conditions for a minimum are used, namely:

$$\frac{\partial \mathcal{L}}{\partial x_1} = 2x_1 - \langle \lambda + 2\rho(4 - x_1 - x_2)\rangle = 0 \tag{3.60}$$

$$\frac{\partial \mathcal{L}}{\partial x_2} = 20x_2 - \langle \lambda + 2\rho(4 - x_1 - x_2)\rangle = 0 \tag{3.61}$$

Clearly

$$x_1 = 10x_2 \tag{3.62}$$

provided that $\langle\ \rangle$ is nonzero. Utilizing (3.60), (3.61) and (3.62), successive iterations can be performed as shown below.

Iteration 1: Use $\lambda^1 = 0$ and $\rho_1 = 1$, then (3.60) and (3.62) imply

$$x_1^{*1} = 1.9047 \text{ and } x_2^{*1} = 0.19047$$

The new λ^2 is now given via (3.59) as $\lambda^2 = \langle \lambda^1 + 2\rho_1(4 - x_1^{*1} - x_2^{*1}) \rangle =$ 3.80952.

Iteration 2: Now with $\lambda^2 = 3.80952$ choose $\rho_2 = 10$ which gives

$x_1^{*2} = 3.2234$, $x_2^{*2} = 0.3223$ and $\lambda^3 = \langle \lambda^2 + 2\rho_2(4 - x_1^{*2} - x_2^{*2}) \rangle = 12.8937$.

Iteration 3: With the above λ^3 and staying with the value of 10 for ρ, the iterations proceed as follows:

Iteration	x_1^{*k}	x_2^{*k}		λ^{k+1}
3	3.5728	0.3573	$\rightarrow$	14.2913
4	3.6266	0.3627	$\rightarrow$	14.5063
5	3.6361	0.3636	$\rightarrow$	14.5861

Thus the iterative procedure converges satisfactorily to the true solution: $\mathbf{x}^* = [3.6363, 0.3636]^T$.

3.5.3 Sequential quadratic programming (SQP)

The SQP method is based on the application of Newton's method to determine $\mathbf{x}^*$ and $\boldsymbol{\lambda}^*$ from the KKT conditions of the constrained optimization problem. It can be shown (see Bazaraa et al., 1993) that the determination of the Newton step is equivalent to the solution of a quadratic programming (QP) problem.

Consider the general problem:

$$\begin{aligned} \text{minimize} \quad & f(\mathbf{x}) \\ \text{such that} \quad & g_j(\mathbf{x}) \leq 0; \; j = 1, 2, \ldots, m \\ & h_j(\mathbf{x}) = 0; \; j = 1, 2, \ldots, r. \end{aligned} \tag{3.63}$$

It *can be shown* (Bazaraa et al., 1993) that given estimates $(\mathbf{x}^k, \boldsymbol{\lambda}^k, \boldsymbol{\mu}^k)$, $k = 0, 1, \ldots$, to the solution and the respective Lagrange multipliers values, with $\boldsymbol{\lambda}^k \geq \mathbf{0}$, then the Newton step $\mathbf{s}$ of iteration $k+1$, such that $\mathbf{x}^{k+1} = \mathbf{x}^k + \mathbf{s}$ is given by the solution to the following *k-th QP problem:*

QP-k $(\mathbf{x}^k, \boldsymbol{\lambda}^k, \boldsymbol{\mu}^k)$: Minimize with respect to $\mathbf{s}$:

$$F(\mathbf{s}) = f(\mathbf{x}^k) + \boldsymbol{\nabla}^T f(\mathbf{x}^k)\mathbf{s} + \tfrac{1}{2}\mathbf{s}^T\mathbf{H}_L(\mathbf{x}^k)\mathbf{s}$$

such that

$$\mathbf{g}(\mathbf{x}^k) + \left[\frac{\partial \mathbf{g}(\mathbf{x}^k)}{\partial \mathbf{x}}\right]^T \mathbf{s} \leq \mathbf{0} \tag{3.64}$$

and

$$\mathbf{h}(\mathbf{x}^k) + \left[\frac{\partial \mathbf{h}(\mathbf{x}^k)}{\partial \mathbf{x}}\right]^T \mathbf{s} = \mathbf{0}$$

and where $\mathbf{g} = [g_1, g_2, \dots, g_m]^T$, $\mathbf{h} = [h_1, h_2, \dots, h_r]^T$ and the Hessian of the classical Lagrangian with respect to $\mathbf{x}$ is

$$\mathbf{H}_L(\mathbf{x}^k) = \boldsymbol{\nabla}^2 f(\mathbf{x}^k) + \sum_{j=1}^{m} \lambda_j^k \boldsymbol{\nabla}^2 g_j(\mathbf{x}^k) + \sum_{j=1}^{r} \mu_j^k \boldsymbol{\nabla}^2 h_j(\mathbf{x}^k).$$

Note that the solution of QP-k does not only yield $\mathbf{s}$, but also the Lagrange multipliers $\boldsymbol{\lambda}^{k+1}$ and $\boldsymbol{\mu}^{k+1}$ via the solution of equation (3.26) in Section 3.4. Thus with $\mathbf{x}^{k+1} = \mathbf{x}^k + \mathbf{s}$ we may construct the next QP problem: QP-$k+1$.

The solution of successive QP problems is continued until $\mathbf{s} = \mathbf{0}$. It can be shown that if this occurs, then the KKT conditions of the original problem (3.63) are satisfied.

In practice, since convergence from a point far from the solution is not guaranteed, the full step $\mathbf{s}$ is usually not taken. To improve convergence, $\mathbf{s}$ is rather used as a search direction in performing a line search minimization of a so-called merit or descent function. A popular choice for the merit function is (Bazaraa et al., 1993):

$$F_E(\mathbf{x}) = f(\mathbf{x}) + \gamma\left(\sum_{i=1}^{m} \max\{0, g_i(\mathbf{x})\} + \sum_{i=1}^{r} |h_i(\mathbf{x})|\right) \tag{3.65}$$

where $\gamma \geq \max\{\lambda_1, \lambda_2, \dots, \lambda_m, |\mu_1|, \dots, |\mu_r|\}$.

Note that it is not advisable here to use curve fitting techniques for the line search since the function is non-differentiable. More stable methods, such as the golden section method, are therefore preferred.

The great advantage of the SQP approach, above the classical Newton method, is that it allows for a systematic and natural way of selecting the active set of constraints and in addition, through the use of the merit function, the convergence process may be controlled.

3.5.3.1 Example

Solve the problem below by means of the basic SQP method, using $\mathbf{x}^0 = [0, 1]^T$ and $\boldsymbol{\lambda}^0 = \mathbf{0}$.

Minimize $f(\mathbf{x}) = 2x_1^2 + 2x_2^2 - 2x_1x_2 - 4x_1 - 6x_2$

such that

$$\begin{aligned} g_1(\mathbf{x}) &= 2x_1^2 - x_2 \leq 0 \\ g_2(\mathbf{x}) &= x_1 + 5x_2 - 5 \leq 0 \\ g_3(\mathbf{x}) &= -x_1 \leq 0 \\ g_4(\mathbf{x}) &= -x_2 \leq 0. \end{aligned}$$

Since $\nabla f(\mathbf{x}) = \begin{bmatrix} 4x_1 - 2x_2 - 4 \\ 4x_2 - 2x_1 - 6 \end{bmatrix}$ it follows that $\nabla f(\mathbf{x}^0) = \begin{bmatrix} -6 \\ -2 \end{bmatrix}$.

Thus with $\boldsymbol{\lambda}^0 = \mathbf{0}$, $\mathbf{H}_L = \begin{bmatrix} 4 & -2 \\ -2 & 4 \end{bmatrix}$ and it follows from (3.64) that the starting quadratic programming problem is:

QP-0: Minimize with respect to $\mathbf{s}$:

$$F(\mathbf{s}) = -4 - 6s_1 - 2s_2 + \tfrac{1}{2}(4s_1^2 + 4s_2^2 - 4s_1s_2)$$

such that

$$-1 - s_2 \leq 0, \qquad s_1 + 5s_2 \leq 0$$
$$-s_1 \leq 0, \ \text{and} \ -1 - s_2 \leq 0$$

where $\mathbf{s} = [s_1, s_2]^T$.

The solution to this QP can be obtained via the method described in Section 3.4, which firstly shows that only the second constraint is active, and then obtains the solution by solving the corresponding equation (3.26) giving $\mathbf{s} = [1.1290, -0.2258]^T$ and $\boldsymbol{\lambda}^1 = [0, 1.0322, 0, 0]^T$ and therefore $\mathbf{x}^1 = \mathbf{x}^0 + \mathbf{s} = [1.1290, 0.7742]^T$ which completes the *first iteration*.

The next quadratic program QP-1 can now be constructed. It is left to the reader to perform the further iterations. The method, because it is basically a Newton method, converges rapidly to the optimum $\mathbf{x}^* = [0.6589, 0.8682]^T$.

Chapter 4

NEW GRADIENT-BASED TRAJECTORY AND APPROXIMATION METHODS

4.1 Introduction

4.1.1 Why new algorithms?

In spite of the mathematical sophistication of classical *gradient-based* algorithms, certain *inhibiting difficulties remain* when these algorithms are applied to *real-world problems.* This is particularly true in the field of engineering, where unique difficulties occur that have prevented the general application of gradient-based mathematical optimization techniques to design problems.

Optimization difficulties that arise are:

(i) the *functions* are often *very expensive to evaluate*, requiring, for example, the time-consuming finite element analysis of a structure, the simulation of the dynamics of a multi-body system, or a computational fluid dynamics (CFD) simulation,

(ii) the *existence of noise*, numerical or experimental, in the functions,

(iii) the presence of *discontinuities* in the functions,

(iv) *multiple local minima*, requiring global optimization techniques,

(v) the existence of regions in the design space where the *functions are not defined*, and

(vi) the occurrence of an *extremely large number* of design *variables*, disqualifying, for example, the SQP method.

4.1.2 Research at the University of Pretoria

All the above difficulties have been addressed in research done at the University of Pretoria over the past twenty years. This research has led to, amongst others, the development of the new optimization algorithms and methods listed in the subsections below.

4.1.2.1 Unconstrained optimization

(i) The *leap-frog* dynamic *trajectory* method: LFOP (Snyman, 1982, 1983),

(ii) a *conjugate-gradient method* with Euler-trapezium steps in which a novel gradient-only line search method is used: ETOP (Snyman, 1985), and

(iii) a *steepest-descent method* applied to successive spherical quadratic approximations: SQSD (Snyman and Hay, 2001).

4.1.2.2 Direct constrained optimization

(i) The leap-frog method for constrained optimization, LFOPC (Snyman, 2000), and

(ii) the conjugate-gradient method with Euler-trapezium steps and gradient-only line searches, applied to penalty function formulations of constrained problems: ETOPC (Snyman, 2003).

4.1.2.3 Approximation methods

(i) A *feasible descent cone method* applied to successive spherical *quadratic sub-problems*: FDC-SAM (Stander and Snyman, 1993; Snyman and Stander, 1994, 1996; De Klerk and Snyman, 1994), and

(ii) the *leap-frog method* (LFOPC) applied to successive spherical *quadratic sub-problems*: Dynamic-Q (Snyman et al., 1994; Snyman and Hay, 2002).

4.1.2.4 Methods for global unconstrained optimization

(i) A *multi-start global minimization* algorithm with *dynamic search trajectories*: SF-GLOB (Snyman and Fatti, 1987), and

(ii) a *modified bouncing ball* trajectory method for global optimization: MBB (Groenwold and Snyman, 2002).

All of the above methods developed at the University of Pretoria are *gradient-based*, and have the common and unique property, for gradient-based methods, that *no explicit objective function line searches are required.*

In this chapter the LFOP/C unconstrained and constrained algorithms are discussed in detail. This is followed by the presentation of the SQSD method, which serves as an introduction to the Dynamic-Q approximation method. Next the ETOP/C algorithms are introduced, with special reference to their ability to deal with the presence of severe noise in the objective function, through the use of a gradient-only line search technique. Finally the SF-GLOB and MBB stochastic global optimization algorithms, which use dynamic search trajectories, are presented and discussed.

4.2 The dynamic trajectory optimization method

The dynamic trajectory method for unconstrained minimization (Snyman, 1982, 1983) is also known as the "leap-frog" method. It has recently (Snyman, 2000) been modified to handle constraints via a penalty function formulation of the constrained problem. The outstanding characteristics of the basic method are:

(i) it uses *only* function *gradient* information ∇f,

(ii) *no* explicit *line searches* are performed,

(iii) it is extremely *robust*, handling steep valleys, and discontinuities and noise in the objective function and its gradient vector, with relative ease,

(iv) the algorithm seeks relative low local minima and can therefore be used as the basic component in a methodology for *global optimization*, and

(v) when applied to smooth and near quadratic functions it is not as efficient as classical methods.

4.2.1 Basic dynamic model

Assume a particle of *unit mass* in a n-dimensional *conservative* force field with *potential energy* at $\mathbf{x}$ given by $f(\mathbf{x})$, then at $\mathbf{x}$ the force (acceleration $\mathbf{a}$) on the particle is given by:

$$\mathbf{a} = \ddot{\mathbf{x}} = -\nabla f(\mathbf{x}) \tag{4.1}$$

from which it follows that for the motion of the particle over the time interval $[0, t]$:

$$\tfrac{1}{2}\|\dot{\mathbf{x}}(t)\|^2 - \tfrac{1}{2}\|\dot{\mathbf{x}}(0)\|^2 = f(\mathbf{x}(0)) - f(\mathbf{x}(t)) \tag{4.2}$$

or

$$T(t) - T(0) = f(0) - f(t) \tag{4.3}$$

where $T(t)$ represents the kinetic energy of the particle at time t. Thus it follows that

$$f(t) + T(t) = \text{constant} \tag{4.4}$$

i.e. conservation of energy along the trajectory. Note that along the particle trajectory the change in the function f, $\Delta f = -\Delta T$, and therefore, as long as T increases, f decreases. This is the underlying principle on which the dynamic leap-frog optimization algorithm is based.

4.2.2 Basic algorithm for unconstrained problems (LFOP)

The basic elements of the LFOP method are as listed in Algorithm 4.1. A detailed flow chart of the basic LFOP algorithm for unconstrained problems is given in Figure 4.1.

4.2.3 Modification for constrained problems (LFOPC)

The code LFOPC (Snyman, 2000) applies the unconstrained optimization algorithm LFOP to a penalty function formulation of the constrained problem (see Section 3.1) in 3 phases. For engineering problems (with convergence tolerance $\varepsilon_x = 10^{-4}$) the choice $\rho_0 = 10$ and $\rho_1 = 100$ is recommended. For extreme accuracy ($\varepsilon_x = 10^{-8}$), use $\rho_0 = 100$ and $\rho_1 = 10^4$.

4.2.3.1 Example

Minimize $f(\mathbf{x}) = x_1^2 + 2x_2^2$ such that $g(\mathbf{x}) = -x_1 - x_2 + 1 \leq 0$ with starting point $\mathbf{x}^0 = [3, 1]^T$ by means of the LFOPC algorithm. Use $\rho_0 = 1.0$ and $\rho_1 = 10.0$. The computed solution is depicted in Figure 4.2.

Algorithm 4.1 LFOP algorithm

1. Given $f(\mathbf{x})$ and starting point $\mathbf{x}(0) = \mathbf{x}^0$, compute the dynamic trajectory of the particle by solving the *initial value problem:*

$$\begin{aligned} \ddot{\mathbf{x}}(t) &= -\nabla f(\mathbf{x}(t)) \\ \text{with } \dot{\mathbf{x}}(0) &= \mathbf{0}; \text{ and } \mathbf{x}(0) = \mathbf{x}^0. \end{aligned} \tag{4.5}$$

2. Monitor $\mathbf{v}(t) = \dot{\mathbf{x}}(t)$, clearly as long as $T = \frac{1}{2}\|\mathbf{v}(t)\|^2$ increases, $f(\mathbf{x}(t))$ decreases as desired.

3. When $\|\mathbf{v}(t)\|$ decreases, i.e. when the particle moves uphill, apply some *interfering strategy* to gradually extract energy from the particle so as to increase the likelihood of its descent, but not so that descent occurs immediately.

4. In practice the numerical integration of the initial value problem (4.5) is done by the *"leap-frog" method:* compute for $k = 0, 1, 2, \ldots$, and given time step Δt:

$$\begin{aligned} \mathbf{x}^{k+1} &= \mathbf{x}^k + \mathbf{v}^k \Delta t \\ \mathbf{v}^{k+1} &= \mathbf{v}^k + \mathbf{a}^{k+1} \Delta t \end{aligned}$$

   where
   $$\mathbf{a}^k = -\nabla f(\mathbf{x}^k) \text{ and } \mathbf{v}^0 = \tfrac{1}{2}\mathbf{a}^0 \Delta t,$$
   to ensure an initial step if $\mathbf{a}^0 \neq \mathbf{0}$.

5. Typically the *interfering strategy* is as follows:
   if $\|\mathbf{v}^{k+1}\| \geq \|\mathbf{v}^k\|$ continue the trajectory,
   else set $\mathbf{v}^k = \frac{1}{4}(\mathbf{v}^{k+1} + \mathbf{v}^k)$, $\mathbf{x}^{k+1} = \frac{1}{2}(\mathbf{x}^{k+1} + \mathbf{x}^k)$, compute new $\mathbf{v}^{k+1}$ and continue.

6. Further *heuristics* (Snyman, 1982, 1983) are introduced to determine a suitable initial time step Δt, to allow for the magnification and reduction of Δt, and to control the magnitude of the step $\Delta\mathbf{x} = \mathbf{x}^{k+1} - \mathbf{x}^k$ by setting a step size limit δ along the computed trajectory. (The recommended magnitude of δ is $\delta \approx \frac{1}{10}\sqrt{n} \times$ (maximum variable range).)

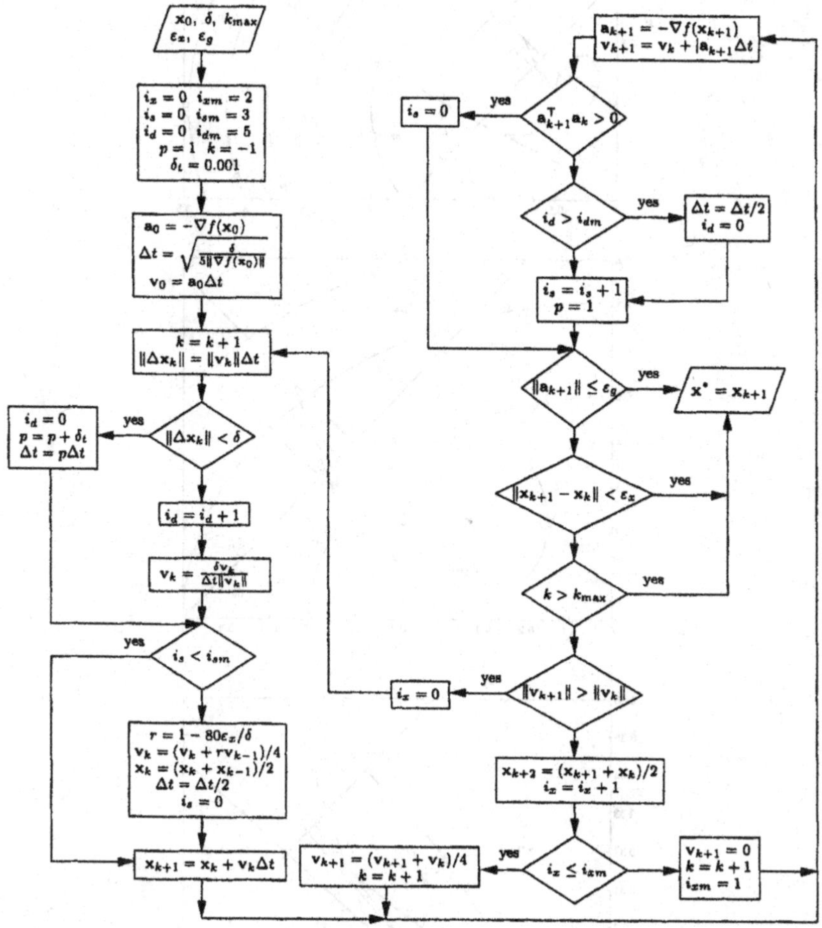

Figure 4.1: Flowchart of the LFOP unconstrained minimization algorithm

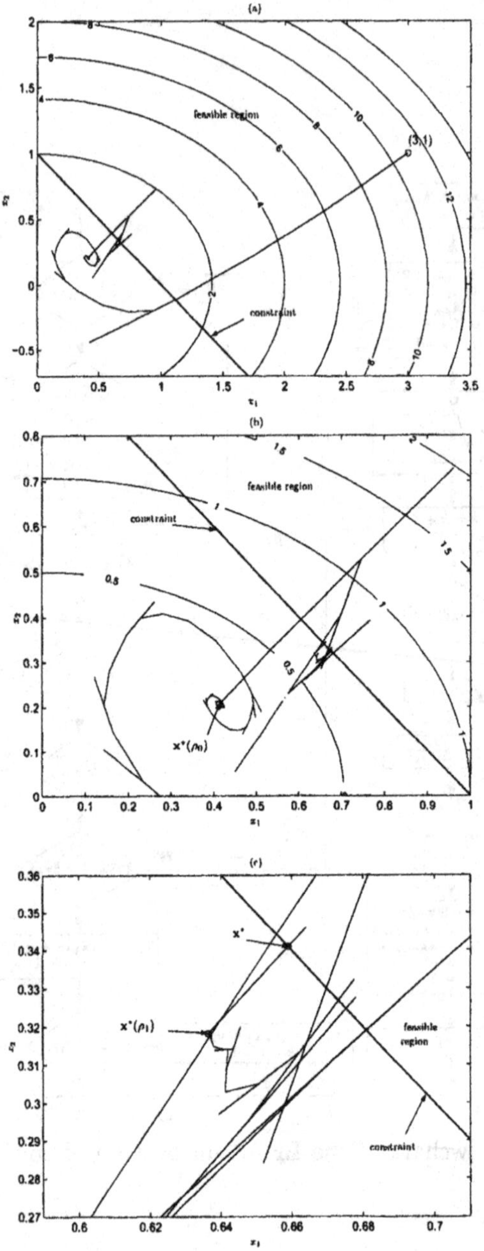

Figure 4.2: The (a) complete LFOPC trajectory for example problem 4.2.3.1, with $\mathbf{x}^0 = [3, 1]^T$, and magnified views of the final part of the trajectory shown in (b) and (c), giving $\mathbf{x}^* \approx [0.659, 0.341]^T$

Algorithm 4.2 LFOPC algorithm

Phase 0:
Given some $\mathbf{x}^0$, then with overall penalty parameter $\rho = \rho_0$, apply LFOP to the penalty function $P(\mathbf{x}, \rho_0)$ to give $\mathbf{x}^*(\rho_0)$.
Phase 1:
With $\mathbf{x}^0 := \mathbf{x}^*(\rho_0)$ and $\rho := \rho_1$, where $\rho_1 \gg \rho_0$, apply LFOP to $P(\mathbf{x}, \rho_1)$ to give $\mathbf{x}^*(\rho_1)$ and identify the set of *active* constraints I_a, such that $g_{i_a}(\mathbf{x}^*(\rho_1)) > 0$ for $i_a \in I_a$.
Phase 2:
With $\mathbf{x}^0 := \mathbf{x}^*(\rho_1)$ use LFOP to minimize

$$P_a(\mathbf{x}, \rho_1) = \sum_{i=1}^{m} \rho_1 h_i^2(\mathbf{x}) + \sum_{i_a \in I_a} \rho_1 g_{i_a}^2(\mathbf{x})$$

to give $\mathbf{x}^*$.

4.3 The spherical quadratic steepest descent method

4.3.1 Introduction

In this section an extremely simple gradient only algorithm (Snyman and Hay, 2001) is proposed that, in terms of storage requirement (only 3 n-vectors need be stored) and computational efficiency, may be considered as an alternative to the conjugate gradient methods. The method effectively applies the steepest descent (SD) method to successive simple spherical quadratic approximations of the objective function in such a way that no explicit line searches are performed in solving the minimization problem. It is shown that the method is convergent when applied to general positive-definite quadratic functions. The method is tested by its application to some standard and other test problems. On the evidence presented the new method, called the SQSD algorithm, appears to be reliable and stable, and very competitive compared to the well established conjugate gradient methods. In particular, it does very well when applied to extremely ill-conditioned problems.

4.3.2 Classical steepest descent method revisited

Consider the following unconstrained optimization problem:

$$\min f(\mathbf{x}),\ \mathbf{x} \in \mathbb{R}^n \tag{4.6}$$

where f is a scalar objective function defined on $\mathbb{R}^n$, the n-dimensional real Euclidean space, and $\mathbf{x}$ is a vector of n real components $x_1, x_2, \ldots, x_n$. It is assumed that f is differentiable so that the gradient vector $\nabla f(\mathbf{x})$ exists everywhere in $\mathbb{R}^n$. The solution is denoted by $\mathbf{x}^*$.

The steepest descent (SD) algorithm for solving problem (4.6) may then be stated as follows:

Algorithm 4.3 SD algorithm

Initialization: Specify convergence tolerances ε_g and ε_x, select starting point $\mathbf{x}^0$. Set $k := 1$ and go to main procedure.
Main procedure:

1. If $\|\nabla f(\mathbf{x}^{k-1})\| < \varepsilon_g$, then set $\mathbf{x}^* \cong \mathbf{x}^c = \mathbf{x}^{k-1}$ and stop; otherwise set $\mathbf{u}^k := -\nabla f(\mathbf{x}^{k-1})$.

2. Let λ_k be such that $f(\mathbf{x}^{k-1} + \lambda_k \mathbf{u}^k) = \min_\lambda f(\mathbf{x}^{k-1} + \lambda \mathbf{u}^k)$ subject to $\lambda \geq 0$ {line search step}.

3. Set $\mathbf{x}^k := \mathbf{x}^{k-1} + \lambda_k \mathbf{u}^k$; if $\|\mathbf{x}^k - \mathbf{x}^{k-1}\| < \varepsilon_x$, then $\mathbf{x}^* \cong \mathbf{x}^c = \mathbf{x}^k$ and stop; otherwise set $k := k + 1$ and go to Step 1.

It can be shown that if the steepest descent method is applied to a general positive-definite quadratic function of the form $f(\mathbf{x}) = \frac{1}{2}\mathbf{x}^T\mathbf{A}\mathbf{x} + \mathbf{b}^T\mathbf{x} + c$, then the sequence $\{f(\mathbf{x}^k)\} \to f(\mathbf{x}^*)$. Depending, however, on the starting point $\mathbf{x}^0$ and the condition number of $\mathbf{A}$ associated with the quadratic form, the rate of convergence may become extremely slow.

It is proposed here that for general functions $f(\mathbf{x})$, better overall performance of the steepest descent method may be obtained by applying it successively to a sequence of very simple quadratic approximations of $f(\mathbf{x})$. The proposed modification, named here the spherical quadratic steepest descent (SQSD) method, remains a first order method since only gradient information is used with no attempt being made to construct the Hessian of the function. The storage requirements therefore remain minimal, making it ideally suitable for problems with a large number of variables. Another significant characteristic is that the method requires no explicit line searches.

4.3.3 The SQSD algorithm

In the SQSD approach, given an initial approximate solution $\mathbf{x}^0$, a sequence of spherically quadratic optimization subproblems $P[k], k = 0, 1, 2, \ldots$ is solved, generating a sequence of approximate solutions $\mathbf{x}^{k+1}$. More specifically, at each point $\mathbf{x}^k$ the constructed approximate subproblem is $P[k]$:

$$\min_{\mathbf{x}} \tilde{f}_k(\mathbf{x}) \tag{4.7}$$

where the approximate objective function $\tilde{f}_k(\mathbf{x})$ is given by

$$\tilde{f}_k(\mathbf{x}) = f(\mathbf{x}^k) + \nabla^T f(\mathbf{x}^k)(\mathbf{x} - \mathbf{x}^k) + \frac{1}{2}(\mathbf{x} - \mathbf{x}^k)^T \mathbf{C}_k (\mathbf{x} - \mathbf{x}^k) \tag{4.8}$$

and $\mathbf{C}_k = \text{diag}(c_k, c_k, \ldots, c_k) = c_k\mathbf{I}$. The solution to this problem will be denoted by $\mathbf{x}^{*k}$, and for the construction of the next subproblem $P[k+1]$, $\mathbf{x}^{k+1} := \mathbf{x}^{*k}$.

For the first subproblem the curvature c_0 is set to $c_0 := \|\nabla f(\mathbf{x}^0)\| / \rho$, where $\rho > 0$ is some arbitrarily specified step limit. Thereafter, for $k \geq 1$, c_k is chosen such that $\tilde{f}(\mathbf{x}^k)$ interpolates $f(\mathbf{x})$ at both $\mathbf{x}^k$ and $\mathbf{x}^{k-1}$. The latter conditions imply that for $k = 1, 2, \ldots$

$$c_k := \frac{2\left[f(\mathbf{x}^{k-1}) - f(\mathbf{x}^k) - \nabla^T f(\mathbf{x}^k)(\mathbf{x}^{k-1} - \mathbf{x}^k)\right]}{\|\mathbf{x}^{k-1} - \mathbf{x}^k\|^2}. \tag{4.9}$$

Clearly the identical curvature entries along the diagonal of the Hessian, mean that the level surfaces of the quadratic approximation $\tilde{f}_k(\mathbf{x})$, are indeed concentric hyper-spheres. The approximate subproblems $P[k]$ are therefore aptly referred to as spherical quadratic approximations.

It is now proposed that for a large class of problems the sequence $\mathbf{x}^0, \mathbf{x}^1, \ldots$ will tend to the solution of the original problem (4.6), i.e.

$$\lim_{k \to \infty} \mathbf{x} = \mathbf{x}^*. \tag{4.10}$$

For subproblems $P[k]$ that are convex, i.e. $c_k > 0$, the solution occurs where $\nabla \tilde{f}_k(\mathbf{x}) = \mathbf{0}$, that is where

$$\nabla f(\mathbf{x}^k) + c_k \mathbf{I}(\mathbf{x} - \mathbf{x}^k) = \mathbf{0}. \tag{4.11}$$

The solution to the subproblem, $\mathbf{x}^{*k}$ is therefore given by

$$\mathbf{x}^{*k} = \mathbf{x}^k - \frac{\nabla f(\mathbf{x}^k)}{c_k}. \tag{4.12}$$

Clearly the solution to the spherical quadratic subproblem lies along a line through $\mathbf{x}^k$ in the direction of steepest descent. The SQSD method may formally be stated in the form given in Algorithm 4.4.

Step size control is introduced in Algorithm 4.4 through the specification of a step limit ρ and the test for $\|\mathbf{x}^k - \mathbf{x}^{k-1}\| > \rho$ in Step 2 of the main procedure. Note that the choice of c_0 ensures that for $P[0]$ the solution $\mathbf{x}^1$ lies at a distance ρ from $\mathbf{x}^0$ in the direction of steepest descent. Also the test in Step 3 that $c_k < 0$, and setting $c_k := 10^{-60}$ where this condition is true ensures that the approximate objective function is always positive-definite.

4.3.4 Convergence of the SQSD method

An analysis of the convergence rate of the SQSD method, when applied to a general positive-definite quadratic function, affords insight into the convergence behavior of the method when applied to more general functions. This is so because for a large class of continuously differentiable functions, the behavior close to local minima is quadratic. For quadratic functions the following theorem may be proved.

4.3.4.1 Theorem

The SQSD algorithm (without step size control) is convergent when applied to the general quadratic function of the form $f(\mathbf{x}) = \frac{1}{2}\mathbf{x}^T\mathbf{A}\mathbf{x} + \mathbf{b}^T\mathbf{x}$, *where* $\mathbf{A}$ *is a* $n \times n$ *positive-definite matrix and* $\mathbf{b} \in \mathbb{R}^n$.

Proof. Begin by considering the bivariate quadratic function, $f(\mathbf{x}) = x_1^2 + \gamma x_2^2$, $\gamma \geq 1$ and with $\mathbf{x}^0 = [\alpha, \beta]^T$. Assume $c_0 > 0$ given, and

Algorithm 4.4 SQSD algorithm

Initialization: Specify convergence tolerances ε_g and ε_x, step limit $\rho > 0$ and select starting point $\mathbf{x}^0$. Set $c_0 := \|\nabla f(\mathbf{x}^0)\| / \rho$. Set $k := 1$ and go to main procedure.

Main procedure:

1. If $\|\nabla f(\mathbf{x}^{k-1})\| < \varepsilon_g$, then $\mathbf{x}^* \cong \mathbf{x}^c = \mathbf{x}^{k-1}$ and stop; otherwise set

$$\mathbf{x}^k := \mathbf{x}^{k-1} - \frac{\nabla f(\mathbf{x}^{k-1})}{c_{k-1}}.$$

2. If $\|\mathbf{x}^k - \mathbf{x}^{k-1}\| > \rho$, then set

$$\mathbf{x}^k := \mathbf{x}^k - \rho \frac{\nabla f(\mathbf{x}^{k-1})}{\|\nabla f(\mathbf{x}^{k-1})\|};$$

   if $\|\mathbf{x}^k - \mathbf{x}^{k-1}\| < \varepsilon_x$, then $\mathbf{x}^* \cong \mathbf{x}^c = \mathbf{x}^k$ and stop.

3. Set

$$c_k := \frac{2\left[f(\mathbf{x}^{k-1}) - f(\mathbf{x}^k) - \nabla^T f(\mathbf{x}^k)(\mathbf{x}^{k-1} - \mathbf{x}^k)\right]}{\|\mathbf{x}^{k-1} - \mathbf{x}^k\|^2};$$

   if $c_k < 0$ set $c_k := 10^{-60}$.

4. Set $k := k + 1$ and go to Step 1 for next iteration.

for convenience in what follows set $c_0 = 1/\delta, \delta > 0$. Also employ the notation $f_k = f(\mathbf{x}^k)$.

Application of the first step of the SQSD algorithm yields

$$\mathbf{x}^1 = \mathbf{x}^0 - \frac{\nabla f_0}{c_0} = [\alpha(1-2\delta), \beta(1-2\gamma\delta)]^T \tag{4.13}$$

and it follows that

$$\|\mathbf{x}^1 - \mathbf{x}^0\|^2 = 4\delta^2(\alpha^2 + \gamma^2\beta^2) \tag{4.14}$$

and

$$\nabla f_1 = [2\alpha(1-2\delta), 2\gamma\beta(1-2\gamma\delta)]^T. \tag{4.15}$$

For the next iteration the curvature is given by

$$c_1 = \frac{2[f_0 - f_1 - \nabla^T f_1(\mathbf{x}^0 - \mathbf{x}^1)]}{\|\mathbf{x}^0 - \mathbf{x}^1\|^2}. \tag{4.16}$$

Utilizing the information contained in (4.13)-(4.15), the various entries in expression (4.16) are known, and after substitution c_1 simplifies to

$$c_1 = \frac{2(\alpha^2 + \gamma^3\beta^2)}{\alpha^2 + \gamma^2\beta^2}. \tag{4.17}$$

In the next iteration, Step 1 gives

$$\mathbf{x}^2 = \mathbf{x}^1 - \frac{\nabla f_1}{c_1}. \tag{4.18}$$

And after the necessary substitutions for $\mathbf{x}^1$, ∇f_1 and c_1, given by (4.13), (4.15) and (4.17) respectively, (4.18) reduces to

$$\mathbf{x}^2 = [\alpha(1-2\delta)\mu_1, \beta(1-2\gamma\delta)\omega_1]^T \tag{4.19}$$

where

$$\mu_1 = 1 - \frac{1 + \gamma^2\beta^2/\alpha^2}{1 + \gamma^3\beta^2/\alpha^2} \tag{4.20}$$

and

$$\omega_1 = 1 - \frac{\gamma + \gamma^3\beta^2/\alpha^2}{1 + \gamma^3\beta^2/\alpha^2}. \tag{4.21}$$

Clearly if $\gamma = 1$, then $\mu_1 = 0$ and $\omega_1 = 0$. Thus by (4.19) $\mathbf{x}^2 = \mathbf{0}$ and convergence to the solution is achieved within the second iteration.

Now for $\gamma > 1$, and for any choice of α and β, it follows from (4.20) that

$$0 \leq \mu_1 \leq 1 \tag{4.22}$$

which implies from (4.19) that for the first component of $\mathbf{x}^2$:

$$\left|x_1^{(2)}\right| = |\alpha(1-2\delta)\mu_1| < |\alpha(1-2\delta)| = \left|x_1^{(1)}\right| \tag{4.23}$$

or introducing α notation (with $\alpha_0 = \alpha$), that

$$|\alpha_2| = |\mu_1\alpha_1| < |\alpha_1|. \tag{4.24}$$

{Note: because $c_0 = 1/\delta > 0$ is chosen arbitrarily, it cannot be said that $|\alpha_1| < |\alpha_0|$. However α_1 is finite.}

The above argument, culminating in result (4.24), is for the two iterations $\mathbf{x}^0 \rightarrow \mathbf{x}^1 \rightarrow \mathbf{x}^2$. Repeating the argument for the sequence of overlapping pairs of iterations $\mathbf{x}^1 \rightarrow \mathbf{x}^2 \rightarrow \mathbf{x}^3$; $\mathbf{x}^2 \rightarrow \mathbf{x}^3 \rightarrow \mathbf{x}^4; \dots$, it follows similarly that $|\alpha_3| = |\mu_2\alpha_2| < |\alpha_2|$; $|\alpha_4| = |\mu_3\alpha_3| < |\alpha_3|; \dots$, since $0 \leq \mu_2 \leq 1$; $0 \leq \mu_3 \leq 1; \dots$, where the μ's are given by (corresponding to equation (4.20) for μ_1):

$$\mu_1 = 1 - \frac{1 + \gamma^2\beta_{j-1}^2/\alpha_{j-1}^2}{1 + \gamma^3\beta_{j-1}^2/\alpha_{j-1}^2}. \tag{4.25}$$

Thus in general

$$0 \leq \mu_j \leq 1 \tag{4.26}$$

and

$$|\alpha_{j+1}| = |\mu_j\alpha_j| < |\alpha_j|. \tag{4.27}$$

For large positive integer m it follows that

$$|\alpha_m| = |\mu_{m-1}\alpha_{m-1}| = |\mu_{m-1}\mu_{m-2}\alpha_{m-2}| = |\mu_{m-1}\mu_{m-2}\cdots\mu_1\alpha_1| \tag{4.28}$$

and clearly for $\gamma > 0$, because of (4.26)

$$\lim_{m\rightarrow\infty} |\alpha_m| = 0. \tag{4.29}$$

Now for the second component of $\mathbf{x}^2$ in (4.19), the expression for ω_1, given by (4.21), may be simplified to

$$\omega_1 = \frac{1-\gamma}{1+\gamma^3\beta^2/\alpha^2}. \tag{4.30}$$

Also for the second component:

$$x_2^{(2)} = \beta(1-2\gamma\delta)\omega_1 = \omega_1 x_2^{(1)} \tag{4.31}$$

or introducing β notation

$$\beta_2 = \omega_1\beta_1. \tag{4.32}$$

The above argument is for $\mathbf{x}^0 \to \mathbf{x}^1 \to \mathbf{x}^2$ and again, repeating it for the sequence of overlapping pairs of iterations, it follows more generally for $j = 1, 2, \ldots$, that

$$\beta_{j+1} = \omega_j\beta_j \tag{4.33}$$

where ω_j is given by

$$\omega_j = \frac{1-\gamma}{1+\gamma^3\beta_{j-1}^2/\alpha_{j-1}^2}. \tag{4.34}$$

Since by (4.29), $|\alpha_m| \to 0$, it follows that if $|\beta_m| \to 0$ as $m \to \infty$, the theorem is proved for the bivariate case. Make the assumption that $|\beta_m|$ does not tend to zero, then there exists a finite positive number ε such that

$$|\beta_j| \geq \varepsilon \tag{4.35}$$

for all j. This allows the following argument:

$$|\omega_j| = \left|\frac{1-\gamma}{1+\gamma^3\beta_{j-1}^2/\alpha_{j-1}^2}\right| \leq \left|\frac{1-\gamma}{1+\gamma^3\varepsilon^2/\alpha_{j-1}^2}\right| = \left|\frac{(1-\gamma)\alpha_{j-1}^2}{\alpha_{j-1}^2+\gamma^3\varepsilon^2}\right|. \tag{4.36}$$

Clearly since by (4.29) $|\alpha_m| \to 0$ as $m \to \infty$, (4.36) implies that also $|\omega_m| \to 0$. This result taken together with (4.33) means that $|\beta_m| \to 0$ which contradicts the assumption above. With this result the theorem is proved for the bivariate case.

Although the algebra becomes more complicated, the above argument can clearly be extended to prove convergence for the multivariate case, where

$$f(\mathbf{x}) = \sum_{i=1}^{n} \gamma_i x_i^2, \ \gamma_1 = 1 < \gamma_2 < \gamma_3 < \ldots < \gamma_n. \quad (4.37)$$

Finally since the general quadratic function

$$f(\mathbf{x}) = \frac{1}{2}\mathbf{x}^T\mathbf{A}\mathbf{x} + \mathbf{b}^T\mathbf{x}, \ \mathbf{A} \text{ positive} - \text{definite} \quad (4.38)$$

may be transformed to the form (4.37), convergence of the SQSD method is also ensured in the general case. □

It is important to note that, although the above analysis indicates that $\|\mathbf{x}^k - \mathbf{x}^*\|$ is monotonically decreasing with k, it does not necessarily follow that monotonic descent in the corresponding objective function values $f(\mathbf{x}^k)$, is also guaranteed. Indeed, numerical experimentation with quadratic functions shows that, although the SQSD algorithm monotonically approaches the minimum, relatively large increases in $f(\mathbf{x}^k)$ may occur along the trajectory to $\mathbf{x}^*$, especially if the function is highly elliptical (poorly scaled).

4.3.5 Numerical results and conclusion

The SQSD method is now demonstrated by its application to some test problems. For comparison purposes the results are also given for the standard SD method and both the Fletcher-Reeves (FR) and Polak-Ribiere (PR) conjugate gradient methods. The latter two methods are implemented using the CG+ FORTRAN conjugate gradient program of Gilbert and Nocedal (1992). The CG+ implementation uses the line search routine of Moré and Thuente (1994). The function and gradient values are evaluated together in a single subroutine. The SD method is applied using CG+ with the search direction modified to the steepest descent direction. The FORTRAN programs were run on a 266 MHz Pentium 2 computer using double precision computations.

The standard (refs. Rao, 1996; Snyman, 1985; Himmelblau, 1972; Manevich, 1999) and other test problems used are listed in Section 4.3.6 and the results are given in Tables 4.1 and 4.2. The convergence tolerances applied throughout are $\varepsilon_g = 10^{-5}$ and $\varepsilon_x = 10^{-8}$, except for the extended homogenous quadratic function with $n = 50000$ (Problem 12) and the extremely ill-conditioned Manevich functions (Problems 14). For these problems the extreme tolerances $\varepsilon_g \cong 0(= 10^{-75})$ and $\varepsilon_x = 10^{-12}$, are prescribed in an effort to ensure very high accuracy in the approximation $\mathbf{x}^c$ to $\mathbf{x}^*$. For each method the number of function-cum-gradient-vector evaluations (N^{fg}) are given. For the SQSD method the number of iterations is the same as N^{fg}. For the other methods the number of iterations (N^{it}) required for convergence, and which corresponds to the number of line searches executed, are also listed separately. In addition the relative error (E^r) in optimum function value, defined by

$$E^r = \left| \frac{f(\mathbf{x}^*) - f(\mathbf{x}^c)}{1 + |f(\mathbf{x}^*)|} \right| \tag{4.39}$$

where $\mathbf{x}^c$ is the approximation to $\mathbf{x}^*$ at convergence, is also listed. For the Manevich problems, with $n \geq 40$, for which the other (SD, FR and PR) algorithms fail to converge after the indicated number of steps, the infinite norm of the error in the solution vector (I^∞), defined by $\|\mathbf{x}^* - \mathbf{x}^c\|_\infty$ is also tabulated. These entries, given instead of the relative error in function value (E^r), are made in italics.

Inspection of the results shows that the SQSD algorithm is consistently competitive with the other three methods and performs notably well for large problems. Of all the methods the SQSD method appears to be the most reliable one in solving each of the posed problems. As expected, because line searches are eliminated and consecutive search directions are no longer forced to be orthogonal, the new method completely overshadows the standard SD method. What is much more gratifying, however, is the performance of the SQSD method relative to the well-established and well-researched conjugate gradient algorithms. Overall the new method appears to be very competitive with respect to computational efficiency and, on the evidence presented, remarkably stable.

In the implementation of the SQSD method to highly non-quadratic and non-convex functions, some care must however be taken in ensuring that the chosen step limit parameter ρ, is not too large. A too large value may result in excessive oscillations occurring before convergence. Therefore a relatively small value, $\rho = 0.3$, was used for the Rosenbrock problem with $n = 2$ (Problem 4). For the extended Rosenbrock functions of

Prob. #	n	SQSD			Steepest Descent		
		ρ	N^{fg}	E^r	N^{fg}	N^{it}	E^r/I^∞
1	3	1	12	3.E-14	41	20	6.E-12
2	2	1	31	1.E-14	266	131	9.E-11
3	2	1	33	3.E-08	2316	1157	4.E-08
4	2	0.3	97	1.E-15	> 20000		3.E-09
5(a)	3	1	11	1.E-12	60	29	6.E-08
5(b)	3	1	17	1.E-12	49	23	6.E-08
6	4	1	119	9.E-09	> 20000		2.E-06
7	3	1	37	1.E-12	156	77	3.E-11
8	2	10	39	1.E-22	12050*	6023*	26*
9	2	0.3	113	5.E-14	6065	3027	2.E-10
10	2	1	43	1.E-12	1309	652	1.E-10
11	4	2	267	2.E-11	16701	8348	4.E-11
12	20	1.E+04	58	1.E-11	276	137	1.E-11
	200	1.E+04	146	4.E-12	2717	1357	1.E-11
	2000	1.E+04	456	2.E-10	> 20000		2.E-08
	20000	1.E+04	1318	6.E-09	> 10000		8.E+01
	50000	1.E+10	4073	3.E-16	> 10000		5.E+02
13	10	0.3	788	2.E-10	> 20000		4.E-07
	100	1	2580	1.E-12	> 20000		3.E+01
	300	1.73	6618	1.E-10	> 20000		2.E+02
	600	2.45	13347	1.E-11	> 20000		5.E+02
	1000	3.16	20717	2.E-10	> 30000		9.E+02
14	20	1	3651	2.E-27	> 20000		9.E-01
		10	3301	9.E-30			
	40	1	13302	5.E-27	> 30000		*1.E+00*
		10	15109	2.E-33			
	60	1	19016	7.E-39	> 30000		*1.E+00*
		10	16023	6.E-39			
	100	1	39690	1.E-49	> 50000		*1.E+00*
		10	38929	3.E-53			
	200	1	73517	5.E-81	> 100000		*1.E+00*
		10	76621	4.E-81			

* Convergence to a local minimum with $f(\mathbf{x}^c) = 48.9$.

Table 4.1: Performance of the SQSD and SD optimization algorithms when applied to the test problems listed in Section 4.3.6

Prob. #	n	Fletcher-Reeves			Polak-Ribiere		
		N^{fg}	N^{it}	E^r/I^∞	N^{fg}	N^{it}	E^r/I^∞
1	3	7	3	0$	7	3	0$
2	2	30	11	2.E-11	22	8	2.E-12
3	2	45	18	2.E-08	36	14	6.E-11
4	2	180	78	1.E-11	66	18	1.E-14
5(a)	3	18	7	6.E-08	18	8	6.E-08
5(b)	3	65	31	6.E-08	26	11	6.E-08
6	4	1573	783	8.E-10	166	68	3.E-09
7	3	132	62	4.E-12	57	26	1.E-12
8	2	72*	27*	26*	24*	11*	26*
9	2	56	18	5.E-11	50	17	1.E-15
10	2	127	60	6.E-12	30	11	1.E-11
11	4	193	91	1.E-12	99	39	9.E-14
12	20	42	20	9.E-32	42	20	4.E-31
	200	163	80	5.E-13	163	80	5.E-13
	2000	530	263	2.E-13	530	263	2.E-13
	20000	1652	825	4.E-13	1652	825	4.E-13
	50000	3225	1161	1.E-20	3225	1611	1.E-20
13	10	> 20000		2.E-02	548	263	4.E-12
	100	> 20000		8.E+01	1571	776	2.E-12
	300	> 20000		3.E+02	3253	1605	2.E-12
	600	> 20000		6.E+02	5550	2765	2.E-12
	1000	> 30000		1.E+03	8735	4358	2.E-12
14	20	187	75	8.E-24	1088	507	2.E-22
	40	> 30000		*1.E+00*	> 30000		*1.E+00*
	60	> 30000		*1.E+00*	> 30000		*1.E+00*
	100	> 50000		*1.E+00*	> 50000		*1.E+00*
	200	> 100000		*1.E+00*	> 100000		*1.E+00*

* Convergence to a local minimum with $f(\mathbf{x}^c) = 48.9$; $ Solution to machine accuracy.

Table 4.2: Performance of the FR and PR algorithms when applied to the test problems listed in Section 4.3.6

larger dimensionality (Problems 13), correspondingly larger step limit values ($\rho = \sqrt{n}/10$) were used with success.

For quadratic functions, as is evident from the convergence analysis of Section 4.3.4, no step limit is required for convergence. This is borne out in practice by the results for the extended homogenous quadratic functions (Problems 12), where the very large value $\rho = 10^4$ was used throughout, with the even more extreme value of $\rho = 10^{10}$ for $n = 50000$. The specification of a step limit in the quadratic case also appears to have little effect on the convergence rate, as can be seen from the results for the ill-conditioned Manevich functions (Problems 14), that are given for both $\rho = 1$ and $\rho = 10$. Here convergence is obtained to at least 11 significant figures accuracy ($\|\mathbf{x}^* - \mathbf{x}^c\|_\infty < 10^{-11}$) for each of the variables, despite the occurrence of extreme condition numbers, such as 10^{60} for the Manevich problem with $n = 200$.

The successful application of the new method to the ill-conditioned Manevich problems, and the analysis of the convergence behavior for quadratic functions, indicate that the SQSD algorithm represents a powerful approach to solving quadratic problems with large numbers of variables. In particular, the SQSD method can be seen as an *unconditionally convergent*, *stable* and *economic* alternative iterative method for solving large systems of linear equations, ill-conditioned or not, through the minimization of the sum of the squares of the residuals of the equations.

4.3.6 Test functions used for SQSD

Minimize $f(\mathbf{x})$:

1. $f(\mathbf{x}) = x_1^2 + 2x_2^2 + 3x_3^2 - 2x_1 - 4x_2 - 6x_3 + 6$, $\mathbf{x}^0 = [3, 3, 3]^T$, $\mathbf{x}^* = [1, 1, 1]^T$, $f(\mathbf{x}^*) = 0.0$.

2. $f(\mathbf{x}) = x_1^4 - 2x_1^2x_2 + x_1^2 + x_2^2 - 2x_1 + 1$, $\mathbf{x}^0 = [3, 3]^T$, $\mathbf{x}^* = [1, 1]^T$, $f(\mathbf{x}^*) = 0.0$.

3. $f(\mathbf{x}) = x_1^4 - 8x_1^3 + 25x_1^2 + 4x_2^2 - 4x_1x_2 - 32x_1 + 16$, $\mathbf{x}^0 = [3, 3]^T$, $\mathbf{x}^* = [2, 1]^T$, $f(\mathbf{x}^*) = 0.0$.

4. $f(\mathbf{x}) = 100(x_2 - x_1^2)^2 + (1 - x_1)^2$, $\mathbf{x}^0 = [-1.2, 1]^T$, $\mathbf{x}^* = [1, 1]^T$, $f(\mathbf{x}^*) = 0.0$ (Rosenbrock's parabolic valley, Rao, 1996).

5. $f(\mathbf{x}) = x_1^4 + x_1^3 - x_1 + x_2^4 - x_2^2 + x_2 + x_3^2 - x_3 + x_1x_2x_3$, (Zlobec's function, Snyman, 1985):

   (a) $\mathbf{x}^0 = [1, -1, 1]^T$ and

   (b) $\mathbf{x}^0 = [0, 0, 0]^T$, $\mathbf{x}^* = [0.57085597, -0.93955591, 0.76817555]^T$, $f(\mathbf{x}^*) = -1.91177218907$.

6. $f(\mathbf{x}) = (x_1 + 10x_2)^2 + 5(x_3 - x_4)^2 + (x_2 - 2x_3)^4 + 10(x_1 - x_4)^4$, $\mathbf{x}^0 = [3, -1, 0, 1]^T$, $\mathbf{x}^* = [0, 0, 0, 0]^T$, $f(\mathbf{x}^*) = 0.0$ (Powell's quartic function, Rao, 1996).

7. $f(\mathbf{x}) = -\left\{\frac{1}{1+(x_1-x_2)^2} + \sin\left(\frac{1}{2}\pi x_2 x_3\right) + \exp\left[-\left(\frac{x_1+x_3}{x_2} - 2\right)^2\right]\right\}$, $\mathbf{x}^0 = [0, 1, 2]^T, \mathbf{x}^* = [1, 1, 1]^T$, $f(\mathbf{x}) = -3.0$ (Rao, 1996).

8. $f(\mathbf{x}) = \{-13 + x_1 + [(5 - x_2)x_2 - 2]x_2\}^2 + \{-29 + x_1 + [(x_2 + 1)x_2 - 14]x_2\}^2$, $\mathbf{x}^0 = [1/2, -2]^T$, $\mathbf{x}^* = [5, 4]^T, f(\mathbf{x}^*) = 0.0$ (Freudenstein and Roth function, Rao, 1996).

9. $f(\mathbf{x}) = 100(x_2 - x_1^3)^2 + (1 - x_1)^2$, $\mathbf{x}^0 = [-1.2, 1]^T$, $\mathbf{x}^* = [1, 1]^T$, $f(\mathbf{x}^*) = 0.0$ (cubic valley, Himmelblau, 1972).

10. $f(\mathbf{x}) = [1.5 - x_1(1 - x_2)]^2 + [2.25 - x_1(1 - x_2^2)]^2 + [2.625 - x_1(1 - x_2^3)]^2$, $\mathbf{x}^0 = [1, 1]^T$, $\mathbf{x}^* = [3, 1/2]^T$, $f(\mathbf{x}^*) = 0.0$ (Beale's function, Rao, 1996).

11. $f(\mathbf{x}) = [10(x_2 - x_1^2)]^2 + (1 - x_1)^2 + 90(x_4 - x_3^2)^2 + (1 - x_3)^2 + 10(x_2 + x_4 - 2)^2 + 0.1(x_2 - x_4)^2$, $\mathbf{x}^0 = [-3, 1, -3, -1]^T$, $\mathbf{x}^* = [1, 1, 1, 1]^T$, $f(\mathbf{x}^*) = 0.0$ (Wood's function, Rao, 1996).

12. $f(\mathbf{x}) = \sum_{i=1}^n ix_i^2$, $\mathbf{x}^0 = [3, 3, \dots, 3]^T$, $\mathbf{x}^* = [0, 0, \dots, 0]^T$, $f(\mathbf{x}^*) = 0.0$ (extended homogeneous quadratic functions).

13. $f(\mathbf{x}) = \sum_{i=1}^{n-1}[100(x_{i+1} - x_i^2)^2 + (1 - x_i)^2]$, $\mathbf{x}^0 = [-1.2, 1, -1.2, 1, \dots]^T$, $\mathbf{x}^* = [1, 1, \dots, 1]^T$, $f(\mathbf{x}^*) = 0.0$ (extended Rosenbrock functions, Rao, 1996).

14. $f(\mathbf{x}) = \sum_{i=1}^n (1 - x_i)^2/2^{i-1}$, $\mathbf{x}^0 = [0, 0, \dots, 0]^T$, $\mathbf{x}^* = [1, 1, \dots, 1]^T$, $f(\mathbf{x}^*) = 0.0$ (extended Manevich functions, Manevich, 1999).

4.4 The Dynamic-Q optimization algorithm

4.4.1 Introduction

An efficient *constrained* optimization method is presented in this section. The method, called the Dynamic-Q method (Snyman and Hay, 2002), consists of applying the dynamic trajectory LFOPC optimization algorithm (see Section 4.2) to successive quadratic approximations of the actual optimization problem. This method may be considered as an extension of the unconstrained SQSD method, presented in Section 4.3, to one capable of handling general constrained optimization problems.

Due to its efficiency with respect to the number of function evaluations required for convergence, the Dynamic-Q method is primarily intended for optimization problems where function evaluations are expensive. Such problems occur frequently in engineering applications where time consuming numerical simulations may be used for function evaluations. Amongst others, these numerical analyses may take the form of a computational fluid dynamics (CFD) simulation, a structural analysis by means of the finite element method (FEM) or a dynamic simulation of a multibody system. Because these simulations are usually expensive to perform, and because the relevant functions may not be known analytically, standard classical optimization methods are normally not suited to these types of problems. Also, as will be shown, the storage requirements of the Dynamic-Q method are minimal. No Hessian information is required. The method is therefore particularly suitable for problems where the number of variables n is large.

4.4.2 The Dynamic-Q method

Consider the general nonlinear optimization problem:

$$\begin{gathered}\min_{\mathbf{x}} f(\mathbf{x});\ \mathbf{x} = [x_1, x_2, \dots, x_n]^T \in \mathbb{R}^n \\ \text{subject to} \\ g_j(\mathbf{x}) = 0;\ j = 1, 2, \dots, p \\ h_k(\mathbf{x}) = 0;\ k = 1, 2, \dots, q\end{gathered} \tag{4.40}$$

where $f(\mathbf{x})$, $g_j(\mathbf{x})$ and $h_k(\mathbf{x})$ are scalar functions of $\mathbf{x}$.

In the Dynamic-Q approach, successive subproblems $P[i]$, $i = 0, 1, 2, \ldots$ are generated, at successive approximations $\mathbf{x}^i$ to the solution $\mathbf{x}^*$, by constructing *spherically quadratic* approximations $\tilde{f}(\mathbf{x})$, $\tilde{g}_j(\mathbf{x})$ and $\tilde{h}_k(\mathbf{x})$ to $f(\mathbf{x})$, $g_j(\mathbf{x})$ and $h_k(\mathbf{x})$. These approximation functions, evaluated at a point $\mathbf{x}^i$, are given by

$$\begin{aligned}
\tilde{f}(\mathbf{x}) &= f(\mathbf{x}^i) + \boldsymbol{\nabla}^T f(\mathbf{x}^i)(\mathbf{x} - \mathbf{x}^i) + \frac{1}{2}(\mathbf{x} - \mathbf{x}^i)^T \mathbf{A}(\mathbf{x} - \mathbf{x}^i) \\
\tilde{g}_j(\mathbf{x}) &= g_j(\mathbf{x}^i) + \boldsymbol{\nabla}^T g_j(\mathbf{x}^i)(\mathbf{x} - \mathbf{x}^i) \\
&\quad + \frac{1}{2}(\mathbf{x} - \mathbf{x}^i)^T \mathbf{B}_j(\mathbf{x} - \mathbf{x}^i), \; j = 1, \ldots, p \qquad (4.41) \\
\tilde{h}_k(\mathbf{x}) &= h_k(\mathbf{x}^i) + \boldsymbol{\nabla}^T h_k(\mathbf{x}^i)(\mathbf{x} - \mathbf{x}^i) \\
&\quad + \frac{1}{2}(\mathbf{x} - \mathbf{x}^i)^T \mathbf{C}_k(\mathbf{x} - \mathbf{x}^i), \; k = 1, \ldots, q
\end{aligned}$$

with the Hessian matrices $\mathbf{A}$, $\mathbf{B}_j$ and $\mathbf{C}_k$ taking on the simple forms

$$\begin{aligned}
\mathbf{A} &= \mathrm{diag}(a, a, \ldots, a) = a\mathbf{I} \\
\mathbf{B}_j &= b_j \mathbf{I} \qquad (4.42) \\
\mathbf{C}_k &= c_k \mathbf{I}.
\end{aligned}$$

Clearly the identical entries along the diagonal of the Hessian matrices indicate that the approximate subproblems $P[i]$ are indeed spherically quadratic.

For the first subproblem ($i = 0$) a linear approximation is formed by setting the curvatures a, b_j and c_k to zero. Thereafter a, b_j and c_k are chosen so that the approximating functions (4.41) interpolate their corresponding actual functions at both $\mathbf{x}^i$ and $\mathbf{x}^{i-1}$. These conditions imply that for $i = 1, 2, 3, \ldots$

$$\begin{aligned}
a &= \frac{2\left[f(\mathbf{x}^{i-1}) - f(\mathbf{x}^i) - \boldsymbol{\nabla}^T f(\mathbf{x}^i)(\mathbf{x}^{i-1} - \mathbf{x}^i)\right]}{\|\mathbf{x}^{i-1} - \mathbf{x}^i\|^2} \qquad (4.43) \\
b_j &= \frac{2\left[g_j(\mathbf{x}^{i-1}) - g_j(\mathbf{x}^i) - \boldsymbol{\nabla}^T g_j(\mathbf{x}^i)(\mathbf{x}^{i-1} - \mathbf{x}^i)\right]}{\|\mathbf{x}^{i-1} - \mathbf{x}^i\|^2}, \; j = 1, \ldots, p \\
c_k &= \frac{2\left[h_k(\mathbf{x}^{i-1}) - h_k(\mathbf{x}^i) - \boldsymbol{\nabla}^T h_k(\mathbf{x}^i)(\mathbf{x}^{i-1} - \mathbf{x}^i)\right]}{\|\mathbf{x}^{i-1} - \mathbf{x}^i\|^2}, \; k = 1, \ldots, q.
\end{aligned}$$

If the gradient vectors $\boldsymbol{\nabla}^T f$, $\boldsymbol{\nabla}^T g_j$ and $\boldsymbol{\nabla}^T h_k$ are not known analytically, they may be approximated from functional data by means of first-order forward finite differences.

The particular choice of spherically quadratic approximations in the Dynamic-Q algorithm has implications on the computational and storage requirements of the method. Since the second derivatives of the objective function and constraints are approximated using function and gradient data, the $O(n^2)$ calculations and storage locations, which would usually be required for these second derivatives, are not needed. The computational and storage resources for the Dynamic-Q method are thus reduced to $O(n)$. At most, $4+p+q+r+s$ n-vectors need be stored (where p, q, r and s are respectively the number of inequality and equality constraints and the number of lower and upper limits of the variables). These savings become significant when the number of variables becomes large. For this reason it is expected that the Dynamic-Q method is well suited, for example, to engineering problems such as structural optimization problems where a large number of variables are present.

In many optimization problems, additional simple side constraints of the form $\hat{k}_i \leq x_i \leq \check{k}_i$ occur. Constants $\hat{k}_i$ and $\check{k}_i$ respectively represent lower and upper bounds for variable x_i. Since these constraints are of a simple form (having zero curvature), they need not be approximated in the Dynamic-Q method and are instead explicitly treated as special linear inequality constraints. Constraints corresponding to lower and upper limits are respectively of the form

$$\begin{aligned} \hat{g}_l(\mathbf{x}) &= \hat{k}_{vl} - x_{vl} \leq 0, \; l = 1, 2, \ldots, r \leq n \\ \check{g}_m(\mathbf{x}) &= x_{wm} - \check{k}_{wm} \leq 0, \; m = 1, 2, \ldots, s \leq n \end{aligned} \tag{4.44}$$

where $vl \in \hat{I} = (v1, v2, \ldots, vr)$ the set of r subscripts corresponding to the set of variables for which respective lower bounds $\hat{k}_{vl}$ are prescribed, and $wm \in \check{I} = (w1, w2, \ldots, ws)$ the set of s subscripts corresponding to the set of variables for which respective upper bounds $\check{k}_{wm}$ are prescribed. The subscripts vl and wm are used since there will, in general, not be n lower and upper limits, i.e. usually $r \neq n$ and $s \neq n$.

In order to obtain convergence to the solution in a controlled and stable manner, move limits are placed on the variables. For each approximate subproblem $P[i]$ this move limit takes the form of an additional single inequality constraint

$$g_\rho(\mathbf{x}) = \left\|\mathbf{x} - \mathbf{x}^{i-1}\right\|^2 - \rho^2 \leq 0 \tag{4.45}$$

where ρ is an appropriately chosen step limit and $\mathbf{x}^{i-1}$ is the solution to the previous subproblem.

The approximate subproblem, constructed at $\mathbf{x}^i$, to the optimization problem (4.41) (plus simple side constraints (4.44) and move limit (4.45)), thus becomes $P[i]$:

$$\begin{aligned}
&\min_{\mathbf{x}} \tilde{f}(\mathbf{x}),\ \mathbf{x} = [x_1, x_2, ..., x_n]^T \in \mathbb{R}^n \\
&\text{subject to} \\
&\tilde{g}_j(\mathbf{x}) \leq 0,\ j = 1, 2, \ldots, p \\
&\tilde{h}_k(\mathbf{x}) = 0,\ k = 1, 2, \ldots, q \\
&\hat{g}_l(\mathbf{x}) \leq 0,\ l = 1, 2, \ldots, r \\
&\check{g}_m(\mathbf{x}) \leq 0,\ m = 1, 2, \ldots, s \\
&g_\rho(\mathbf{x}) = \left\|\mathbf{x} - \mathbf{x}^{i-1}\right\|^2 - \rho^2 \leq 0
\end{aligned} \tag{4.46}$$

with solution $\mathbf{x}^{*i}$. The Dynamic-Q algorithm is given by Algorithm 4.5. In the Dynamic-Q method the subproblems generated are solved using the dynamic trajectory, or "leap-frog" (LFOPC) method of Snyman (1982, 1983) for unconstrained optimization applied to penalty function formulations (Snyman et al., 1994; Snyman, 2000) of the constrained problem. A brief description of the LFOPC algorithm is given in Section 4.2.

Algorithm 4.5 Dynamic-Q algorithm

Initialization: Select starting point $\mathbf{x}^0$ and move limit ρ. Set $i := 0$.
Main procedure:

1. Evaluate $f(\mathbf{x}^i)$, $g_j(\mathbf{x}^i)$ and $h_k(\mathbf{x}^i)$ as well as $\nabla f(\mathbf{x}^i)$, $\nabla g_j(\mathbf{x}^i)$ and $\nabla h_k(\mathbf{x}^i)$. If termination criteria are satisfied set $\mathbf{x}^* = \mathbf{x}^i$ and stop.

2. Construct a local approximation $P[i]$ to the optimization problem at $\mathbf{x}^i$ using expressions (4.41) to (4.43).

3. Solve the approximated subproblem $P[i]$ (given by (4.46)) using the constrained optimizer LFOPC with $\mathbf{x}^0 := \mathbf{x}^i$ (see Section 4.2) to give $\mathbf{x}^{*i}$.

4. Set $i := i + 1$, $\mathbf{x}^i := \mathbf{x}^{*(i-1)}$ and return to Step 2.

The LFOPC algorithm possesses a number of outstanding characteristics, which makes it highly suitable for implementation in the Dynamic-Q methodology. The algorithm requires only gradient information and no explicit line searches or function evaluations are performed. These properties, together with the influence of the fundamental physical principles underlying the method, ensure that the algorithm is extremely robust. This has been proven over many years of testing (Snyman, 2000). A further desirable characteristic related to its robustness, and the main reason for its application in solving the subproblems in the Dynamic-Q algorithm, is that if there is no feasible solution to the problem, the LFOPC algorithm will still find the best possible compromised solution without breaking down. The Dynamic-Q algorithm thus usually converges to a solution from an infeasible remote point without the need to use line searches between subproblems, as is the case with SQP. The LFOPC algorithm used by Dynamic-Q is identical to that presented in Snyman (2000) except for a minor change to LFOP which is advisable should the subproblems become effectively unconstrained.

4.4.3 Numerical results and conclusion

The Dynamic-Q method requires very few parameter settings by the user. Other than convergence criteria and specification of a maximum number of iterations, the only parameter required is the step limit ρ. The algorithm is not very sensitive to the choice of this parameter, however, ρ should be chosen of the same order of magnitude as the diameter of the region of interest. For the problems listed in Table 4.3 a step limit of $\rho = 1$ was used except for problems 72 and 106 where step limits and $\rho = 100$ were used respectively.

Given specified positive tolerances ε_x, ε_f and ε_c, then at step i termination of the algorithm occurs if the normalized step size

$$\frac{\|\mathbf{x}^i - \mathbf{x}^{i-1}\|}{1 + \|\mathbf{x}^i\|} < \varepsilon_x \tag{4.47}$$

or if the normalized change in function value

$$\frac{|f^i - f_{\text{best}}|}{1 + |f_{\text{best}}|} < \varepsilon_f \tag{4.48}$$

where f_{best} is the lowest previous feasible function value and the current $\mathbf{x}^i$ is feasible. The point $\mathbf{x}^i$ is considered feasible if the absolute value of the violation of each constraint is less than ε_c. This particular function termination criterion is used since the Dynamic-Q algorithm may at times exhibit oscillatory behavior near the solution.

In Table 4.3, for the same starting points, the performance of the Dynamic-Q method on some standard test problems is compared to results obtained for Powell's SQP method as reported by Hock and Schittkowski (1981). The problem numbers given correspond to the problem numbers in Hock and Schittkowski's book. For each problem, the actual function value f_{act} is given, as well as, for each method, the calculated function value f^* at convergence, the relative function error

$$E^r = \frac{|f_{\text{act}} - f^*|}{1 + |f_{\text{act}}|} \tag{4.49}$$

and the number of function-gradient evaluations (N^{fg}) required for convergence. In some cases it was not possible to calculate the relative function error due to rounding off of the solutions reported by Hock and Schittkowski. In these cases the calculated solutions were correct to at least eight significant figures. For the Dynamic-Q algorithm, convergence tolerances of $\varepsilon_f = 10^{-8}$ on the function value, $\varepsilon_x = 10^{-5}$ on the step size and $\varepsilon_c = 10^{-6}$ for constraint feasibility, were used. These were chosen to allow for comparison with the reported SQP results.

The result for the 12-corner polytope problem of Svanberg (1999) is also given. For this problem the results given in the SQP columns are for Svanberg's Method of Moving Asymptotes (MMA). The recorded number of function evaluations for this method is approximate since the results given correspond to 50 outer iterations of the MMA, each requiring about 3 function evaluations.

A robust and efficient method for nonlinear optimization, with minimal storage requirements compared to those of the SQP method, has been proposed and tested. The particular methodology proposed is made possible by the special properties of the LFOPC optimization algorithm (Snyman, 2000), which is used to solve the quadratic subproblems. Comparison of the results for Dynamic-Q with the results for the SQP method show that equally accurate results are obtained with comparable number of function evaluations.

Prob. #	n	f_{act}	SQP			Dynamic-Q		
			N^{fg}	f^*	E^r	N^{fg}	f^*	E^r
2	2	5.04E-02	16~	2.84E+01	2.70E+01	7*	4.94E+00	<1.00E-08
10	2	-1.00E+00	12	-1.00E+00	5.00E-08	13	-1.00E+00	<1.00E-08
12	2	-3.00E+01	12	-3.00E+01	<1.00E-08	9	-3.00E+01	<1.00E-08
13	2	1.00E+00	45	1.00E+00	5.00E-08	50$	9.59E-01	2.07E-02
14	2	1.39E+00	6	1.39E+00	8.07E-09	5	1.39E+00	7.86E-07
15	2	3.07E+02	5	3.07E+02	<1.00E-08	15*	3.60E+02	5.55E-07
16	2	2.50E-01	6*	2.31E+01	<1.00E-08	5*	2.31E+01	<1.00E-08
17	2	1.00E+00	12	1.00E+00	<1.00E-08	16	1.00E+00	<1.00E-08
20	2	3.82E+01	20	3.82E+01	4.83E-09	4*	4.02E+01	<1.00E-08
22	2	1.00E+00	9	1.00E+00	<1.00E-08	3	1.00E+00	<1.00E-08
23	2	2.00E+00	7	2.00E+00	<1.00E-08	5	2.00E+00	<1.00E-08
24	2	-1.00E+00	5	-1.00E+00	<1.00E-08	4	-1.00E+00	1.00E-08
26	3	0.00E+00	19	4.05E-08	4.05E-08	27	1.79E-07	1.79E-07
27	3	4.00E-02	25	4.00E-02	1.73E-08	28	4.00E-02	9.62E-10
28	3	0.00E+00	5	2.98E-21	2.98E-21	12	7.56E-10	7.56E-10
29	3	-2.26E+01	13	-2.26E+01	8.59E-11	11	-2.26E+01	8.59E-11
30	3	1.00E+00	14	1.00E+00	<1.00E-08	5	1.00E+00	<1.00E-08
31	3	6.00E+00	10	6.00E+00	<1.00E-08	10	6.00E+00	1.43E-08
32	3	1.00E+00	3	1.00E+00	<1.00E-08	4	1.00E+00	<1.00E-08
33	3	-4.59E+00	5*	-4.00E+00	<1.00E-08	3*	-4.00E+00	<1.00E-08
36	3	-3.30E+03	4	-3.30E+03	<1.00E-08	15	-3.30E+03	<1.00E-08
45	5	1.00E+00	8	1.00E+00	<1.00E-08	7	1.00E+00	1.00E-08
52	5	5.33E+00	8	5.33E+00	5.62E-09	12	5.33E+00	1.02E-08
55	6	6.33E+00	1~	6.00E+00	4.54E-02	2*	6.66E+00	1.30E-09
56	7	-3.46E+00	11	-3.46E+00	<1.00E-08	20	-3.46E+00	6.73E-08
60	3	3.26E-02	9	3.26E-02	3.17E-08	11	3.26E-02	1.21E-09
61	3	-1.44E+02	10	-1.44E+02	1.52E-08	10	-1.44E+02	1.52E-08
63	3	9.62E+02	9	9.62E+02	2.18E-09	6	9.62E+02	2.18E-09
65	3	9.54E-01	11~	2.80E+00	9.47E-01	9	9.54E-01	2.90E-08
71	4	1.70E+01	5	1.70E+01	1.67E-08	6	1.70E+01	1.67E-08
72	4	7.28E+02	35	7.28E+02	1.37E-08	30	7.28E+02	1.37E-08
76	4	-4.68E+00	6	-4.68E+00	3.34E-09	8	-4.68E+00	3.34E-09
78	5	-2.92E+00	9	-2.92E+00	2.55E-09	6	-2.92E+00	2.55E-09
80	5	5.39E-02	7	5.39E-02	7.59E-10	6	5.39E-02	7.59E-10
81	5	5.39E-02	8	5.39E-02	1.71E-09	12	5.39E-02	1.90E-10
100	7	6.80E+02	20	6.80E+02	<1.00E-08	16	6.80E+02	1.46E-10
104	8	3.95E+00	19	3.95E+00	8.00E-09	42	3.95E+00	5.26E-08
106	8	7.05E+03	44	7.05E+03	1.18E-05	79	7.05E+03	1.18E-05
108	9	-8.66E-01	9*	-6.97E-01	1.32E-02	26	-8.66E-01	3.32E-09
118	15	6.65E+02	~	~	~	38	6.65E+02	3.00E-08
Svan	21	2.80E+02	150	2.80E+02	9.96E-05	93	2.80E+02	1.59E-06

*** Converges to a local minimum - listed E^r relative to function value at local minimum;**
~ Fails; $ Terminates on maximum number of steps

Table 4.3: Performance of the Dynamic-Q and SQP optimization algorithms

4.5 A gradient-only line search method for conjugate gradient methods

4.5.1 Introduction

Many engineering design optimization problems involve numerical computer analyses via, for example, FEM codes, CFD simulations or the computer modeling of the dynamics of multi-body mechanical systems. The computed objective function is therefore often the result of a complex sequence of calculations involving other computed or measured quantities. This may result in the presence of numerical noise in the objective function so that it exhibits non-smooth trends as design parameters are varied. It is well known that this presence of numerical noise in the design optimization problem inhibits the use of classical and traditional gradient-based optimization methods that employ line searches, such as for example, the conjugate gradient methods. The numerical noise may prevent or slow down convergence during optimization. It may also promote convergence to spurious local optima. The computational expense of the analyses, coupled to the convergence difficulties created by the numerical noise, is in many cases a significant obstacle to performing multidisciplinary design optimization.

In addition to the problems anticipated when applying the conjugate gradient methods to noisy functions, it is also known that standard implementations of conjugate gradient methods, in which conventional line search techniques have been used, are less robust than one would expect from their the theoretical quadratic termination property. Therefore the conjugate gradient method would, under normal circumstances, not be preferred to quasi-Newton methods (Fletcher, 1987). In particular severe numerical difficulties arise when standard line searches are used in solving constrained problems through the minimization of associated penalty functions. However, there is one particular advantage of conjugate gradient methods, namely the particular simple form that requires no matrix operations in determining the successive search directions. Thus, conjugate gradient methods may be the only methods which are applicable to large problems with thousands of variables (Fletcher, 1987), and are therefore well worth further investigation.

In this section a new implementation (ETOPC) of the conjugate gradient method (both for the Fletcher-Reeves and Polak-Ribiere versions (see Fletcher, 1987) is presented for solving constrained problem. The essential novelty in this implementation is the use of a gradient-only line search technique originally proposed by the author (Snyman, 1985), and used in the ETOP algorithm for unconstrained minimization. It will be shown that this implementation of the conjugate gradient method, not only easily overcomes the accuracy problem when applied to the minimization of penalty functions, but also economically handles the problem of severe numerical noise superimposed on an otherwise smooth underlying objective function.

4.5.2 Formulation of optimization problem

Consider again the general constrained optimization problem:

$$\min_{\mathbf{x}} f(\mathbf{x}),\ \mathbf{x} = [x_1, x_2, x_3, \ldots, x_n]^T \in R^n \tag{4.50}$$

subject to the inequality and equality constraints:

$$\begin{aligned} g_j(\mathbf{x}) &\leq 0,\ j = 1, 2, \ldots, m \\ h_j(\mathbf{x}) &= 0,\ j = 1, 2, \ldots, r \end{aligned} \tag{4.51}$$

where the objective function $f(\mathbf{x})$, and the constraint functions $g_j(\mathbf{x})$ and $h_j(\mathbf{x})$, are scalar functions of the real column vector $\mathbf{x}$. The optimum solution is denoted by $\mathbf{x}^*$, with corresponding optimum function value $f(\mathbf{x}^*)$.

The most straightforward way of handling the constraints is via the unconstrained minimization of the penalty function:

$$P(\mathbf{x}) = f(\mathbf{x}) + \sum_{j=1}^{r} \rho_j h_j^2(\mathbf{x}) + \sum_{j=1}^{m} \beta_j g_j^2(\mathbf{x}) \tag{4.52}$$

where $\rho_j \gg 0$, $\beta_j = 0$ if $g_j(\mathbf{x}) \leq 0$, and $\beta_j = \mu_j \gg 0$ if $g_j(\mathbf{x}) > 0$.

Usually $\rho_j = \mu_j = \mu \gg 0$ for all j, with the corresponding penalty function being denoted by $P(\mathbf{x}, \mu)$.

Central to the application of the conjugate gradient method to penalty function formulated problems presented here, is the use of an unconventional line search method for unconstrained minimization, proposed by the author, in which no function values are explicitly required (Snyman, 1985). Originally this gradient-only line search method was applied to the conjugate gradient method in solving a few very simple unconstrained problems. For somewhat obscure reasons, given in the original paper (Snyman, 1985) and briefly hinted to in this section, the combined method (novel line search plus conjugate gradient method) was called the ETOP (Euler-Trapezium Optimizer) algorithm. For this historical reason, and to avoid confusion, this acronym will be retained here to denote the combined method for unconstrained minimization. In subsequent unreported numerical experiments, the author was successful in solving a number of more challenging practical constrained optimization problems via penalty function formulations of the constrained problem, with ETOP being used in the unconstrained minimization of the sequence of penalty functions. ETOP, applied in this way to constrained problems, was referred to as the ETOPC algorithm. Accordingly this acronym will also be used here.

4.5.3 Gradient-only line search

The line search method used here, and originally proposed by the author (Snyman, 1985) uses no explicit function values. Instead the line search is implicitly done by using only two gradient vector evaluations at two points along the search direction and assuming that the function is near-quadratic along this line. The essentials of the gradient-only line search, for the case where the function $f(\mathbf{x})$ is unconstrained, are as follows. Given the current design point $\mathbf{x}^k$ at iteration k and next search direction $\mathbf{v}^{k+1}$, then compute

$$\mathbf{x}^{k+1} = \mathbf{x}^k + \mathbf{v}^{k+1}\tau \tag{4.53}$$

where τ is some suitably chosen positive parameter. The step taken in (4.53) may be seen as an "Euler step". With this step given by

$$\Delta\mathbf{x}^k = \mathbf{x}^{k+1} - \mathbf{x}^k = \mathbf{v}^{k+1}\tau \tag{4.54}$$

the line search in the direction $\mathbf{v}^{k+1}$ is equivalent to finding $\mathbf{x}^{*k+1}$ defined by

$$f(\mathbf{x}^{*k+1}) = \min_{\lambda} f(\mathbf{x}^k + \lambda\Delta\mathbf{x}^k). \tag{4.55}$$

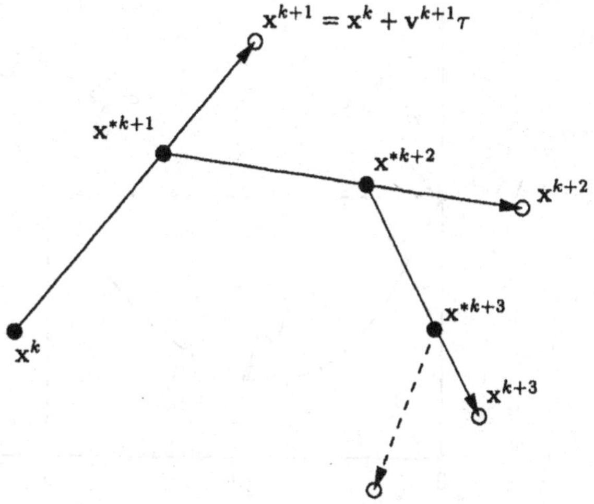

Figure 4.3: Successive steps in line search procedure

These steps are depicted in Figure 4.3.

It was indicated in Snyman (1985) that for the step $\mathbf{x}^{k+1} = \mathbf{x}^k + \mathbf{v}^{k+1}\tau$ the change in function value Δf_k, in the unconstrained case, can be approximated without explicitly evaluating the function $f(\mathbf{x})$. Here a more formal argument is presented via the following lemma.

4.5.3.1 Lemma 1

For a general quadratic function the change in function value, for the step $\Delta\mathbf{x}^k = \mathbf{x}^{k+1} - \mathbf{x}^k = \mathbf{v}^{k+1}\tau$ is given by:

$$\Delta f_k = -\langle \mathbf{v}^{k+1}, \frac{1}{2}(\mathbf{a}^k + \mathbf{a}^{k+1})\tau \rangle \quad (4.56)$$

where $\mathbf{a}^k = -\nabla f(\mathbf{x}^k)$ and $\langle\ ,\ \rangle$ denotes the scalar product.

Proof:

In general, by Taylor's theorem:

$$f(\mathbf{x}^{k+1}) - f(\mathbf{x}^k) = \langle \mathbf{x}^{k+1} - \mathbf{x}^k, \nabla f(\mathbf{x}^k) \rangle + \frac{1}{2}\langle \Delta\mathbf{x}^k, \mathbf{H}(\mathbf{x}^a)\Delta\mathbf{x}^k \rangle$$

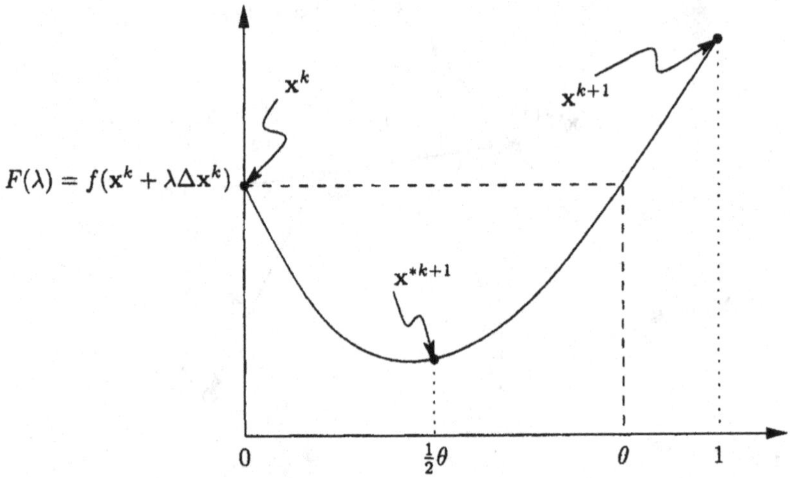

Figure 4.4: Approximation of minimizer $\mathbf{x}^{*k+1}$ in the direction $\mathbf{v}^{k+1}$

and

$$f(\mathbf{x}^{k+1}) - f(\mathbf{x}^k) = \langle \mathbf{x}^{k+1} - \mathbf{x}^k, \nabla f(\mathbf{x}^{k+1})\rangle - \frac{1}{2}\langle \Delta\mathbf{x}^k, \mathbf{H}(\mathbf{x}^b)\Delta\mathbf{x}^k\rangle$$

where $\mathbf{x}^a = \mathbf{x}^k + \theta_0\Delta\mathbf{x}^k$, $\mathbf{x}^b = \mathbf{x}^k + \theta_1\Delta\mathbf{x}^k$ and both θ_0 and θ_1 in the interval $[0,1]$, and where $\mathbf{H}(\mathbf{x})$ denotes the Hessian matrix of the general function $f(\mathbf{x})$. Adding the above two expressions gives:

$$\begin{aligned} f(\mathbf{x}^{k+1}) - f(\mathbf{x}^k) &= \frac{1}{2}\langle \mathbf{x}^{k+1} - \mathbf{x}^k, \nabla f(\mathbf{x}^k) + \nabla f(\mathbf{x}^{k+1})\rangle \\ &\quad + \frac{1}{4}\langle \Delta\mathbf{x}^k, [\mathbf{H}(\mathbf{x}^a) - \mathbf{H}(\mathbf{x}^b)]\Delta\mathbf{x}^k\rangle. \end{aligned}$$

If $f(\mathbf{x})$ is quadratic then $\mathbf{H}(\mathbf{x})$ is constant and it follows that

$$\Delta f_k = f(\mathbf{x}^{k+1}) - f(\mathbf{x}^k) = -\langle \mathbf{v}^{k+1}, \frac{1}{2}(\mathbf{a}^k + \mathbf{a}^{k+1})_T\rangle$$

where $\mathbf{a}^k = \nabla f(\mathbf{x}^k)$, which completes the proof. □

By using expression (4.56) the position of the minimizer $\mathbf{x}^{*k+1}$ (see Figure 4.4), in the direction $\mathbf{v}^{k+1}$, can also be approximated without any explicit function evaluation. This conclusion follows formally from the second lemma given below. Note that in (4.56) the second quantity in the scalar product corresponds to an average vector given by the "trapezium rule". This observation together with the remark following equation (4.53), gave rise to the name "Euler-trapezium optimizer (ETOP)" when applying this line search technique in the conjugate gradient method.

4.5.3.2 Lemma 2

For $f(\mathbf{x})$ a positive-definite quadratic function the point $\mathbf{x}^{*k+1}$ defined by $f(\mathbf{x}^{*k+1}) = \min_\lambda f(\mathbf{x}^k + \lambda\Delta\mathbf{x}^k)$ is given by

$$\mathbf{x}^{*k+1} = \mathbf{x}^k + \frac{1}{2}\theta\Delta\mathbf{x}^k \tag{4.57}$$

where

$$\theta = \rho/(\langle \mathbf{v}^{k+1}, \frac{1}{2}(\mathbf{a}^k + \mathbf{a}^{k+1})\tau\rangle + \rho) \text{ and } \rho = -\langle\Delta\mathbf{x}^k, \mathbf{a}^k\rangle. \tag{4.58}$$

Proof:

First determine θ such that

$$f(\mathbf{x}^k + \theta\Delta\mathbf{x}^k) = f(\mathbf{x}^k).$$

By Taylor's expansion:

$$f(\mathbf{x}^{k+1}) - f(\mathbf{x}^k) = \rho + \frac{1}{2}\langle\Delta\mathbf{x}^k, \mathbf{H}\Delta\mathbf{x}^k\rangle, \text{ i.e., } \frac{1}{2}\langle\Delta\mathbf{x}^k, \mathbf{H}\Delta\mathbf{x}^k\rangle = \Delta f_k - \rho$$

which gives for the step $\theta\Delta\mathbf{x}$:

$$f(\mathbf{x}^k + \theta\Delta\mathbf{x}^k) - f(\mathbf{x}^k) = \theta\rho + \frac{1}{2}\theta^2\langle\Delta\mathbf{x}^k, \mathbf{H}\Delta\mathbf{x}^k\rangle = \theta\rho + \theta^2(\Delta f_k - \rho).$$

For both function values to be the same, θ must therefore satisfy:

$$0 = \theta(\rho + \theta(\Delta f_k - \rho))$$

which has the non-trivial solution:

$$\theta = -\rho/(\Delta f_k - \rho).$$

Using the expression for Δf_k given by Lemma 1, it follows that:

$$\theta = \rho/(\langle\mathbf{v}^{k+1}, \frac{1}{2}(\mathbf{a}^k + \mathbf{a}^{k+1})\tau\rangle + \rho)$$

and by the symmetry of quadratic functions that

$$\mathbf{x}^{*k+1} = \mathbf{x}^k + \frac{1}{2}\theta\Delta\mathbf{x}^k.$$

□

Expressions (4.57) and (4.58) may of course also be used in the general non-quadratic case, to determine an approximation to the minimizer $\mathbf{x}^{*k+1}$ in the direction $\mathbf{v}^{k+1}$, when performing successive line searches using the sequence of descent directions, $\mathbf{v}^{k+1}$, $k = 1, 2, \ldots$ Thus in practice, for the next $(k+1)$-th iteration, set $\mathbf{x}^{k+1} := \mathbf{x}^{*k+1}$, and with the next selected search direction $\mathbf{v}^{k+2}$ proceed as above, using expressions (4.57) and (4.58) to find $\mathbf{x}^{*k+2}$ and then set $\mathbf{x}^{k+2} := \mathbf{x}^{*k+2}$. Continue iterations in this way, with only two gradient vector evaluations done per line search, until convergence is obtained.

In summary, explicit function evaluations are unnecessary in the above line search procedure, since the two computed gradients along the search direction allow for the computation of an approximation (4.56) to the change in objective function, which in turn allows for the estimation of the position of the minimum along the search line via expressions (4.57) and (4.58), based on the assumption that the function is near quadratic in the region of the search.

4.5.3.3 Heuristics

Of course in general the objective function may not be quadratic and positive-definite. Additional heuristics are therefore required to ensure descent, and to see to it that the step size (corresponding to the parameter τ between successive gradient evaluations, is neither too small nor too large. The details of these heuristics are as set out below.

(i) In the case of a successful step having been taken, with Δf_k computed via (4.56) negative, i.e. descent, and θ computed via (4.58) positive, i.e. the function is locally strictly convex, as shown in Figure 4.4, τ is increased by a factor of 1.5 for the next search direction.

(ii) It may turn out that although Δf_k computed via (4.56) is negative, that θ computed via (4.58) is also negative. The latter implies that the function along the search direction is locally concave. In this case set $\theta := -\theta$ in computing $\mathbf{x}^{*k+1}$ by (4.57), so as to ensure a step in the descent direction, and also increase τ by the factor 1.5 before computing the step for the next search direction using (4.53).

(iii) It may happen that Δf_k computed by (4.56) is negative and exactly equal to ρ, i.e. $\Delta f_k - \rho = 0$. This implies zero curvature with $\theta = \infty$ and the function is therefore locally linear. In this case enforce the value $\theta = 1$. This results in the setting, by (4.57), of $\mathbf{x}^{*k+1}$ equal to a point halfway between $\mathbf{x}^k$ and $\mathbf{x}^{k+1}$. In this case τ is again increased by the factor of 1.5.

(iv) If both Δf_k and θ are positive, which is the situation depicted in Figure 4.4, then τ is halved before the next step.

(v) In the only outstanding and unlikely case, should it occur, where Δf^k is positive and θ negative, τ is unchanged.

(vi) For usual unconstrained minimization the initial step size parameter selection is $\tau = 0.5$.

The new gradient-only line search method may of course be applied to any line search descent method for the unconstrained minimization of a general multi-variable function. Here its application is restricted to the conjugate gradient method.

4.5.4 Conjugate gradient search directions and SUMT

The search vectors used here correspond to the conjugate gradient directions (Bazaraa et al., 1993). In particular for $k = 0, 1, \ldots$, the search vectors are

$$\mathbf{v}^{k+1} = (-\nabla f(\mathbf{x}^k) + \beta_{k+1}\mathbf{v}^k/\tau)\tau = \mathbf{s}^{k+1}\tau \tag{4.59}$$

where $\mathbf{s}^{k+1}$ denote the usual conjugate gradient directions, $\beta_1 = 0$ and for $k > 0$:

$$\beta_{k+1} = \|\nabla f(\mathbf{x}^k)\|^2/\|\nabla f(\mathbf{x}^{k-1})\|^2 \tag{4.60}$$

for the Fletcher-Reeves implementation, and for the Polak-Ribiere version:

$$\beta_{k+1} = \langle \nabla f(\mathbf{x}^k) - \nabla f(\mathbf{x}^{k-1}), \nabla f(\mathbf{x}^k) \rangle / \|\nabla f(\mathbf{x}^{k-1})\|^2. \tag{4.61}$$

As recommended by Fletcher (1987), the conjugate gradient algorithm is restarted in the direction of steepest descent when $k > n$.

For the constrained problem the unconstrained minimization is of course applied to successive penalty function formulations $P(\mathbf{x})$ of the form shown in (4.52), using the well known Sequential Unconstrained Minimization Technique (SUMT) (Fiacco and McCormick, 1968). In SUMT, for $j = 1, 2, \ldots$, until convergence, successive unconstrained minimizations are performed on successive penalty functions $P(\mathbf{x}) = P(\mathbf{x}, \mu^{(j)})$ in which the overall penalty parameter $\mu^{(j)}$ is successively increased: $\mu^{(j+1)} := 10\mu^{(j)}$. The corresponding initial step size parameter is set at $\tau = 0.5/\mu^{(j)}$ for each sub problem j. This application of ETOP to the constrained problem, via the unconstrained minimization of successive penalty functions, is referred to as the ETOPC algorithm. In practice, if analytical expressions for the components of the gradient of the objective function are not available, they may be calculated with sufficient accuracy by finite differences. However, when the presence of severe noise is suspected, the application of the gradient-only search method with conjugate gradient search directions, requires that *central finite difference* approximations to the gradients be used in order to effectively smooth out the noise. In this case relatively excessive perturbations δx_i in x_i must be used, which in practice may typically be of the order of 0.1 times the range of interest!

In the application of ETOPC a limit Δ_m, is in practice set to the maximum allowable magnitude Δ^* of the step $\boldsymbol{\Delta}^* = \mathbf{x}^{*k+1} - \mathbf{x}^k$. If Δ^* is greater than Δ_m, then set

$$\mathbf{x}^{k+1} := \mathbf{x}^k + (\mathbf{x}^{*k+1} - \mathbf{x}^k)\Delta_m/\Delta^* \tag{4.62}$$

and restart the conjugate gradient procedure, with $\mathbf{x}^0 := \mathbf{x}^{k+1}$, in the direction of steepest descent. If the maximum allowable step is taken n times in succession, then Δ_m is doubled.

4.5.5 Numerical results

The proposed new implementation of the conjugate gradient method (both the Fletcher- Reeves and Polak-Ribiere versions) is tested here using 40 different problems arbitrarily selected from the famous set of test problems of Hock and Schittkowski (1981). The problem numbers (Pr. #) in the tables, correspond to the numbering used in Hock and Schittkowski. The final test problem, (12-poly), is the 12 polytope problem of Svanberg (1995, 1999). The number of variables (n) of the test problems ranges from 2 to 21 and the number of constraints (m plus r) per problem, from 1 to 59. The termination criteria for the ETOPC algorithm are as follows:

(i) Convergence tolerances for successive approximate sub-problems within SUMT: ε_g for convergence on the norm of the gradient vector, i.e. terminate if $\|\nabla P(\mathbf{x}^{*k+1}, \mu)\| < \varepsilon_g$, and ε_x for convergence on average change of design vector: i.e. terminate if $\frac{1}{2}\|\mathbf{x}^{*k+1} - \mathbf{x}^{*k-1}\| < \varepsilon_x$.

(ii) Termination of the SUMT procedure occurs if the absolute value of the relative difference between the objective function values at the solution points of successive SUMT problems is less than ε_f.

4.5.5.1 Results for smooth functions with no noise

For the initial tests no noise is introduced. For high accuracy requirements (relative error in optimum objective function value to be less than 10^{-8}), it is found that the proposed new conjugate gradient implementation performs as robust as, and more economical than, the traditional penalty function implementation, FMIN, of Kraft and Lootsma reported in Hock and Schittkowski (1981). The detailed results are as tabulated in Table 4.4. Unless otherwise indicated the algorithm settings are: $\varepsilon_x = 10^{-8}$, $\varepsilon_g = 10^{-5}$, $\Delta_m = 1.0$, $\varepsilon_f = 10^{-8}$, $\mu^{(0)} = 1.0$ and $iout = 15$, where $iout$ denotes the maximum number of SUMT iterations allowed. The number of gradient vector evaluations required by ETOPC for the different problems are denoted by nge (note that the number of explicit function evaluations is zero), and the relative error in function value at convergence to the point $\mathbf{x}^c$ is denoted by r_f, which is computed from

$$r_f = |f(\mathbf{x}^*) - f(\mathbf{x}^c)|/(|f(\mathbf{x}^*)| + 1). \tag{4.63}$$

For the FMIN algorithm only the number of explicit objective function evaluations nfe are listed, together with the relative error r_f at convergence. The latter method requires, in addition to the number of function evaluations listed, a comparable number of gradient vector evaluations, which is not given here (see Hock and Schittkowski, 1981).

4.5.5.2 Results for severe noise introduced in the objective function

Following the successful implementation for the test problems with no noise, all the tests were rerun, but with severe relative random noise introduced in the objective function $f(\mathbf{x})$ and all gradient components computed by central finite differences. The influence of noise is investigated for two cases, namely, for a variation of the superimposed uniformly distributed random noise as large as (i) 5% and (ii) 10% of $(1 + |f(\mathbf{x}^*)|)$, where $\mathbf{x}^*$ is the optimum of the underlying smooth problem. The detailed results are shown in Table 4.5. The results are listed only for the Fletcher-Reeves version. The results for the Polak- Ribiere implementation are almost identical. Unless otherwise indicated the algorithm settings are: $\delta x_i = 1.0$, $\varepsilon_g = 10^{-5}$, $\Delta_m = 1.0$, $\varepsilon_f = 10^{-8}$, $\mu^{(0)} = 1.0$ and $iout = 6$, where $iout$ denotes the maximum number of SUMT iterations allowed. For termination of sub-problem on step size, ε_x was set to $\varepsilon_x := 0.005\sqrt{n}$ for the initial sub-problem. Thereafter it is successively halved for each subsequent sub-problem.

The results obtained are surprisingly good with, in most cases, fast convergence to the neighbourhood of the known optimum of the underlying smooth problem. In 90% of the cases regional convergence was obtained with relative errors $r_x < 0.025$ for 5% noise and $r_x < 0.05$ for 10% noise, where

$$r_x = \|\mathbf{x}^* - \mathbf{x}^c\|/(\|\mathbf{x}^*\| + 1) \tag{4.64}$$

and $\mathbf{x}^c$ denotes the point of convergence. Also in 90% of the test problems the respective relative errors in final objective function values were $r_f < 0.025$ for 5% noise and $r_f < 0.05$ for 10% noise, where r_f is as defined in (4.63).

		Fletcher-Reeves		Polak-Ribiere		FMIN	
Pr. #	$n\ m\ r$	nge	r_f	nge	r_f	nfe	r_f
1	2 1 -	100	$< 10^{-14}$	103	$< 10^{-13}$	549	$< 10^{-8}$
2	2 1 -	290	$< 10^{-8}$	318	$< 10^{-8}$	382	1×10^{-8}
10	2 1 -	231	$< 10^{-9}$	247	$< 10^{-9}$	289	7×10^{-8}
12	2 1 -	163	$< 10^{-10}$	184	$< 10^{-10}$	117	1×10^{-8}
13[1]	2 3 -	4993[2]	0.028	4996[2]	0.034	1522	0.163
14	2 1 1	214	$< 10^{-10}$	200	$< 10^{-10}$	232	2×10^{-7}
15	2 3 -	699	$< 10^{-9}$	632	$< 10^{-9}$	729	4×10^{-7}
16	2 5 -	334	$< 10^{-9}$	284	$< 10^{-7}$	362	1×10^{-8}
17	2 5 -	218	$< 10^{-9}$	209	$< 10^{-9}$	541	1×10^{-8}
20[3]	2 5 -	362	$< 10^{-9}$	375	$< 10^{-9}$	701	4×10^{-6}
22	2 2 -	155	$< 10^{-9}$	202	$< 10^{-9}$	174	1×10^{-7}
23	2 9 -	257	$< 10^{-9}$	244	$< 10^{-9}$	423	6×10^{-6}
24	2 5 -	95	$< 10^{-11}$	163	2×10^{-6}	280	2×10^{-8}
26	3 - 1	78	$< 10^{-8}$	100	2×10^{-8}	182	1×10^{-8}
27	3 - 1	129	$< 10^{-8}$	115	$< 10^{-8}$	173	1×10^{-8}
28	3 - 1	17	$< 10^{-28}$	17	$< 10^{-28}$	23	$< 10^{-8}$
29	3 1 -	254	$< 10^{-10}$	267	$< 10^{-10}$	159	$< 10^{-8}$
30	3 7 -	115	$< 10^{-10}$	124	$< 10^{-10}$	1199	4×10^{-8}
31	3 7 1	309	$< 10^{-9}$	274	$< 10^{-9}$	576	$< 10^{-8}$
32	3 7 1	205	$< 10^{-10}$	207	$< 10^{-10}$	874	$< 10^{-8}$
33[3]	3 6 -	272[3]	$< 10^{-10}$	180	$< 10^{-10}$	672[3]	3×10^{-7}
36	3 7 -	336	$< 10^{-12}$	351	$< 10^{-10}$	263	2×10^{-6}
45	5 10 -	175	$< 10^{-10}$	150	$< 10^{-10}$	369	$< 10^{-8}$
52	5 - 3	403	$< 10^{-9}$	388	$< 10^{-9}$	374	$< 10^{-8}$
55[3]	6 8 6	506	$< 10^{-9}$	488	$< 10^{-9}$	581[3]	3×10^{-8}
56	7 - 4	316	6×10^{-8}	289	7×10^{-8}	446	$< 10 - 8$
60	3 6 1	198	$< 10^{-10}$	189	$< 10^{-10}$	347	$1 \times 10 - 8$
61	3 - 2	205	$< 10^{-10}$	201	$< 10^{-10}$	217	$< 10 - 8$
63	3 3 2	205	$< 10^{-10}$	208	$< 10^{-10}$	298	$< 10 - 8$
65	3 7 -	179	$< 10^{-8}$	198	$< 10^{-10}$	-	fails
71	4 9 1	493	$< 10^{-9}$	536	$< 10^{-9}$	1846	5×10^{-3}
72[4]	4 10 -	317	$< 10^{-10}$	298	$< 10^{-10}$	1606	5×10^{-2}
76	4 7 -	224	$< 10^{-10}$	227	$< 10^{-10}$	424	$< 10^{-8}$
78	5 - 3	261	$< 10^{-10}$	264	$< 10^{-10}$	278	$< 10^{-8}$
80	5 10 3	192	$< 10^{-11}$	194	$< 10^{-11}$	1032	2×10^{-8}
81[5]	5 10 3	138	$< 10^{-11}$	158	$< 10^{-10}$	1662	5×10^{-7}
106	8 22 -	6060	5×10^{-6}	6496	3×10^{-5}	-	fails
108	9 14 -	600	$< 10^{-10}$	519	$< 10^{-10}$	984	7×10^{-5}
118[6]	15 29 -	1233	$< 10^{-8}$	1358	$< 10^{-8}$	-	fails
12-poly[6]	21 22 -	844	$< 10^{-9}$	1478	$< 10^{-9}$	-	-

[1]Constraint qualification not satisfied. [2]Termination on maximum number of steps.
[3]Convergence to local minimum. [4]$\mu^{(0)} = 1.0$, $\Delta_m = 1.0$. [5]$\mu^{(0)} = 10^2$.
[6]Gradients by central finite differences, $\delta x_i = 10^{-6}$, $\varepsilon_x = 10^{-6}$.

Table 4.4: The respective performances of the new conjugate gradient implementation ETOPC and FMIN for test problems with no noise introduced

		5% noise			10% noise		
Pr. #	nmr	nge	r_f	r_x	nge	r_f	r_x
1	2 1 -	54	0.035	5×10^{-3}	54	0.06	5×10^{-3}
2	2 1 -	80	2×10^{-3}	2×10^{-3}	87	9×10^{-3}	2×10^{-3}
10	2 1 -	120	0.02	0.022	160	0.048	0.023
12	2 1 -	99	0.018	8×10^{-2}	232	0.006	0.011
13	2 3 -	394	0.079	0.044	187	0.189	0.095
14	2 1 1	138	0.025	6×10^{-4}	126	0.041	6×10^{-4}
15	2 3 -	152	6×10^{-5}	2×10^{-5}	154	0.006	8×10^{-5}
16[1]	2 5 -	250	0.128	0.13	175	0.135	0.16
17	2 5 -	84	0.012	7×10^{-6}	77	0.041	3×10^{-4}
20	2 5 -	89	0.009	2×10^{-5}	105	0.001	2×10^{-5}
22	2 2 -	75	0.01	4×10^{-5}	86	0.035	9×10^{-5}
23	2 9 -	103	0.008	9×10^{-4}	100	0.005	7×10^{-4}
24	2 5 -	75	0.0095	4×10^{-5}	137	0.014	3×10^{-5}
26	3 - 1	63	0.019	2×10^{-3}	71	0.04	3×10^{-3}
27	3 - 1	159	0.015	0.014	132	0.022	0.036
28	3 - 1	46	0.018	6×10^{-3}	49	0.009	0.025
29	3 1 -	232	0.013	0.01	251	0.046	0.015
30	3 7 -	52	0.025	4×10^{-3}	72	0.043	6×10^{-3}
31	3 7 1	123	0.015	9×10^{-4}	183	0.031	0.013
32	3 7 1	89	0.006	4×10^{-3}	107	0.031	5×10^{-3}
33	3 6 -	183	0.016	0.035	83	0.026	3×10^{-3}
36[2]	3 7 -	177	0.018	6×10^{-5}	179	0.01	8×10^{-5}
45	5 10 -	122	0.0013	9×10^{-4}	92	0.009	4×10^{-5}
52	5 - 3	239	0.019	0.042	318	0.041	0.071
55	6 8 6	137	0.016	5×10^{-3}	188	0.041	4×10^{-3}
56	7 - 4	166	0.012	0.014	144	0.03	0.038
60	3 6 1	95	0.021	0.071	83	0.018	0.033
61[2]	3 - 2	105	0.019	2×10^{-3}	83	0.026	9×10^{-4}
63[2]	3 3 2	198	0.02	8×10^{-3}	652	0.016	0.06
65	3 7 -	94	4×10^{-3}	3×10^{-3}	106	0.012	0.003
71	4 9 1	164	0.021	0.022	143	0.021	0.035
72	4 10 -	454	0.01	0.025	578	0.005	0.094
76	4 7 -	131	0.022	0.002	148	0.012	0.041
78	5 - 3	87	0.011	0.004	88	0.037	0.002
80	5 10 3	92	0.011	0.025	105	0.005	0.02
81[3]	5 10 3	39	0.017	0.032	47	0.012	0.031
106[4]	8 22 -	6016	0.023	0.088	8504	0.038	0.113
108[2]	9 14 -	113	0.017	0.04	140	0.04	0.025
118[2]	15 29 -	395	0.012	0.041	371	0.049	0.1
12-poly[2]	21 22 -	476	0.012	0.065	607	0.047	0.1

[1] $\delta x_i = 10^{-1}$. [2] $\delta x_i = 10$. [3] $\mu^{(0)} = 10^2$ [4] $\delta x_i = 10^3$. [5] $\Delta_m = 10^2$.

Table 4.5: Performance of ETOPC for test problems with severe noise introduced

4.5.6 Conclusion

The ETOPC algorithm performs exceptionally well for a first order method in solving constrained problems where the functions are smooth. For these problems the gradient only penalty function implementation of the conjugate gradient method performs as well, if not better than the best conventional implementations reported in the literature, in producing highly accurate solutions.

In the cases where severe noise is introduced in the objective function, relatively fast convergence to the neighborhood of $\mathbf{x}^*$, the solution of the underlying smooth problem, is obtained. Of interest is the fact that with the reduced accuracy requirement associated with the presence of noise, the number of function evaluations required to obtain sufficiently accurate solutions in the case of noise, is on the average much less than that necessary for the high accuracy solutions for smooth functions. As already stated, ETOPC yields in 90% of the cases regional convergence with relative errors $r_x < 0.025$ for 5% noise, and $r_x < 0.05$ for 10% noise. Also in 90% of the test problems the respective relative errors in the final objective function values are $r_f < 0.025$ for 5% noise and $r_f < 0.05$ for 10% noise. In the other 10% of the cases the relative errors are also acceptably small. These accuracies are more than sufficient for multidisciplinary design optimization problems where similar noise may be encountered.

4.6 Global optimization using dynamic search trajectories

4.6.1 Introduction

The problem of globally optimizing a real valued function is inherently intractable (unless hard restrictions are imposed on the objective function) in that no practically useful characterization of the global optimum is available. Indeed the problem of determining an accurate estimate of the global optimum is mathematically ill-posed in the sense that very similar objective functions may have global optima very distant from each other (Schoen, 1991). Nevertheless, the need in practice to find a relative low local minimum has resulted in considerable research over the last decade to develop algorithms that attempt to find such a low minimum, e.g. see Törn and Zilinskas (1989).

The general global optimization problem may be formulated as follows. Given a real valued objective function $f(\mathbf{x})$ defined on the set $\mathbf{x} \in D$ in $\mathbb{R}^n$, find the point $\mathbf{x}^*$ and the corresponding function value f^* such that

$$f^* = f(\mathbf{x}^*) = \text{ minimum } \{f(\mathbf{x})|\mathbf{x} \in D\} \tag{4.65}$$

if such a point $\mathbf{x}^*$ exists. If the objective function and/or the feasible domain D are non-convex, then there may be many local minima which are not global.

If D corresponds to all $\mathbb{R}^n$ the optimization problem is *unconstrained*. Alternatively, simple bounds may be imposed, with D now corresponding to the hyper box (or domain or region of interest) defined by

$$D = \{\mathbf{x}|\boldsymbol{\ell} \leq \mathbf{x} \leq \boldsymbol{u}\} \tag{4.66}$$

where $\boldsymbol{\ell}$ and $\boldsymbol{u}$ are n-vectors defining the respective lower and upper bounds on $\mathbf{x}$.

From a *mathematical* point of view, Problem (4.65) is essentially *unsolvable*, due to a lack of mathematical conditions characterizing the global optimum, as opposed to the local optimum of a smooth continuous function, which is characterized by the behavior of the problem function (Hessians and gradients) at the minimum (Arora et al., 1995) (viz. the Karush-Kuhn-Tucker conditions). Therefore, the global optimum f^* can only be obtained by an exhaustive search, except if the objective function satisfies certain subsidiary conditions (Griewank, 1981), which mostly are of limited practical use (Snyman and Fatti, 1987). Typically, the conditions are that f should satisfy a Lipschitz condition with known constant L and that the search area is bounded, e.g. for all $\mathbf{x}, \bar{\mathbf{x}} \in \mathbf{X}$

$$|f(\mathbf{x}) - f(\bar{\mathbf{x}})| \leq L\|\mathbf{x} - \bar{\mathbf{x}}\|. \tag{4.67}$$

So called space-covering deterministic techniques have been developed (Dixon et al., 1975) under these special conditions. These techniques are expensive, and due to the need to know L, of limited practical use.

Global optimization algorithms are divided into two major classes (Dixon et al., 1975): deterministic and stochastic (from the Greek word *stokhastikos*, i.e. 'governed by the laws of probability'). Deterministic methods can be used to determine the global optimum through exhaustive search. These methods are typically extremely expensive. With the introduction of a stochastic element into deterministic algorithms, the deterministic *guarantee* that the global optimum can be found is relaxed into a *confidence measure*. Stochastic methods can be used to assess the probability of having obtained the global minimum. Stochastic ideas are mostly used for the development of stopping criteria, or to approximate the regions of attraction as used by some methods (Arora et al., 1995).

The stochastic algorithms presented herein, namely the Snyman-Fatti algorithm and the modified bouncing ball algorithm (Groenwold and Snyman, 2002), both depend on dynamic search trajectories to minimize the objective function. The respective trajectories, namely the motion of a particle of unit mass in a n-dimensional conservative force field, and the trajectory of a projectile in a conservative gravitational field, are modified to increase the likelihood of convergence to a low local minimum.

4.6.2 The Snyman-Fatti trajectory method

The essentials of the original SF algorithm (Snyman and Fatti, 1987) using dynamic search trajectories for unconstrained global minimization will now be discussed. The algorithm is based on the local algorithms presented by Snyman (1982, 1983). For more details concerning the motivation of the method, its detailed construction, convergence theorems, computational aspects and some of the more obscure heuristics employed, the reader is referred to the original paper.

4.6.2.1 Dynamic trajectories

In the SF algorithm successive sample points $\mathbf{x}^j, j = 1, 2, ...,$ are selected at random from the box D defined by (4.66). For *each* sample point $\mathbf{x}^j$, a sequence of trajectories T^i, $i = 1, 2, ...,$ is computed by numerically solving the successive initial value problems:

$$\ddot{\mathbf{x}}(t) = -\nabla f(\mathbf{x}(t))$$
$$\mathbf{x}(0) = \mathbf{x}_0^i\ ;\ \ \dot{\mathbf{x}}(0) = \dot{\mathbf{x}}_0^i. \tag{4.68}$$

This trajectory represents the motion of a particle of unit mass in a n-dimensional conservative force field, where the function to be minimized represents the potential energy.

Trajectory T^i is terminated when $\mathbf{x}(t)$ reaches a point where $f(\mathbf{x}(t))$ is arbitrarily close to the value $f(\mathbf{x}_0^i)$ while moving "uphill", or more precisely, if $\mathbf{x}(t)$ satisfies the conditions

$$f(\mathbf{x}(t)) > f(\mathbf{x}_0^i) - \epsilon_u$$
$$\text{and} \quad \dot{\mathbf{x}}(t)^T \nabla f(\mathbf{x}(t)) > 0 \tag{4.69}$$

where ϵ_u is an arbitrary small prescribed positive value.

An argument is presented in Snyman and Fatti (1987) to show that when the level set $\{\mathbf{x}|f(\mathbf{x}) \leq f(\mathbf{x}_0^i)\}$ is bounded and $\nabla f(\mathbf{x}_0^i) \neq \mathbf{0}$, then conditions (4.69) above will be satisfied at some finite point in time.

Each computed step along trajectory T^i is monitored so that at termination the point $\mathbf{x}_m^i$ at which the minimum value was achieved is recorded together with the associated velocity $\dot{\mathbf{x}}_m^i$ and function value f_m^i. The values of $\mathbf{x}_m^i$ and $\dot{\mathbf{x}}_m^i$ are used to determine the initial values for the next trajectory T^{i+1}. From a comparison of the minimum values the best point $\mathbf{x}_b^i$, for the current j over all trajectories to date is also recorded. In more detail the minimization procedure for *a given sample point* $\mathbf{x}^j$, in computing the sequence $\mathbf{x}_b^i$, $i = 1, 2, ...$, is as follows.

In the original paper (Snyman and Fatti, 1987) an argument is presented to indicate that under normal conditions on the continuity of f and its derivatives, $\mathbf{x}_b^i$ will converge to a local minimum. Procedure *MP1*, for a given j, is accordingly terminated at step 3 above if $||\nabla f(\mathbf{x}_b^i)|| \leq \epsilon$, for some small prescribed positive value ϵ, and $\mathbf{x}_b^i$ is taken as the local minimizer $\mathbf{x}_f^j$, i.e. set $\mathbf{x}_f^j := \mathbf{x}_b^i$ with corresponding function value $f_f^j := f(\mathbf{x}_f^j)$.

Reflecting on the overall approach outlined above, involving the computation of energy conserving trajectories and the minimization procedure,

Algorithm 4.6 Minimization Procedure *MP1*

1. For given sample point $\mathbf{x}^j$, set $\mathbf{x}_0^1 := \mathbf{x}^j$ and compute T^1 subject to $\dot{\mathbf{x}}_0^1 := 0$; record $\mathbf{x}_m^1, \dot{\mathbf{x}}_m^1$ and f_m^1 ; set $\mathbf{x}_b^1 := \mathbf{x}_m^1$ and $i := 2$,
2. compute trajectory T^i with $\mathbf{x}_0^i := \frac{1}{2}\left(\mathbf{x}_0^{i-1} + \mathbf{x}_b^{i-1}\right)$ and $\dot{\mathbf{x}}_0^i := \frac{1}{2}\dot{\mathbf{x}}_m^{i-1}$, record $\mathbf{x}_m^i, \dot{\mathbf{x}}_m^i$ and f_m^i,
3. if $f_m^i < f(\mathbf{x}_b^{i-1})$ then $\mathbf{x}_b^i := \mathbf{x}_m^i$; else $\mathbf{x}_b^i := \mathbf{x}_b^{i-1}$,
4. set $i := i + 1$ and go to 2.

it should be evident that, in the presence of many local minima, the probability of convergence to a relative low local minimum is increased. This one expects because, with a small value of ϵ_u (see conditions (4.69)), it is likely that the particle will move through a trough associated with a relative high local minimum, and move over a ridge to record a lower function value at a point beyond. Since we assume that the level set associated with the starting point function is bounded, termination of the search trajectory will occur as the particle eventually moves to a region of higher function values.

4.6.3 The modified bouncing ball trajectory method

The essentials of the modified bouncing ball algorithm using dynamic search trajectories for unconstrained global minimization are now presented. The algorithm is in an experimental stage, and details concerning the motivation of the method, its detailed construction, and computational aspects will be presented in future.

4.6.3.1 Dynamic trajectories

In the MBB algorithm successive sample points $\mathbf{x}^j, j = 1, 2, ...,$ are selected at random from the box D defined by (4.66). For *each* sample point $\mathbf{x}^j$, a sequence of *trajectory steps* $\Delta\mathbf{x}^i$ and associated *projection points* $\mathbf{x}^{i+1}$, $i = 1, 2, ...,$ are computed from the successive analytical relationships (with $\mathbf{x}^1 := \mathbf{x}^j$ and prescribed $V_{0_1} > 0$):

$$\Delta\mathbf{x}^i = V_{0_i} t_i \cos\theta_i \nabla f(\mathbf{x}^i)/\|\nabla f(\mathbf{x}^i)\| \tag{4.70}$$

where

$$\theta_i = \tan^{-1}(\|\nabla f(\mathbf{x}^i)\|) + \frac{\pi}{2}, \tag{4.71}$$

$$t_i = \frac{1}{g}\left[V_{0_i}\sin\theta_i + \left\{(V_{0_i}\sin\theta_i)^2 + 2gh(\mathbf{x}^i)\right\}^{1/2}\right], \tag{4.72}$$

$$h(\mathbf{x}^i) = f(\mathbf{x}^i) + k \tag{4.73}$$

with k a constant chosen such that $h(\mathbf{x}) > 0 \ \forall\ \mathbf{x} \in D$, g a positive constant, and

$$\mathbf{x}^{i+1} = \mathbf{x}^i + \Delta\mathbf{x}^i. \tag{4.74}$$

For the next step, select $V_{0_{i+1}} < V_{0_i}$. Each step $\Delta\mathbf{x}^i$ represents the ground or horizontal displacement obtained by projecting a particle in a vertical gravitational field (constant g) at an elevation $h(\mathbf{x}^i)$ and speed V_{0_i} at an inclination θ_i. The angle θ_i represents the angle that the outward normal $\mathbf{n}$ to the hypersurface represented by $y = h(\mathbf{x})$ makes, at $\mathbf{x}^i$ in $n+1$ dimensional space, with the horizontal. The time of flight t_i is the time taken to reach the ground corresponding to $y = 0$.

More formally, the minimization trajectory for *a given sample point* $\mathbf{x}^j$ and some initial prescribed speed V_0 is obtained by computing the sequence $\mathbf{x}^i$, $i = 1, 2, \ldots$, as follows.

Algorithm 4.7 Minimization Procedure *MP2*

1. For given sample point $\mathbf{x}^j$, set $\mathbf{x}^1 := \mathbf{x}^j$ and compute trajectory step $\Delta\mathbf{x}^1$ according to (4.70) - (4.73) and subject to $V_{0_1} := V_0$; record $\mathbf{x}^2 := \mathbf{x}^1 + \Delta\mathbf{x}^1$, set $i := 2$ and $V_{0_2} := \alpha V_{0_1}$ ($\alpha < 1$).

2. Compute $\Delta\mathbf{x}^i$ according to (4.70) - (4.73) to give $\mathbf{x}^{i+1} := \mathbf{x}^i + \Delta\mathbf{x}^i$, record $\mathbf{x}^{i+1}$ and set $V_{0_{i+1}} := \alpha V_{0_i}$.

3. Set $i := i + 1$ and go to 2.

In the vicinity of a local minimum $\hat{\mathbf{x}}$ the sequence of projection points $\mathbf{x}^i$, $i = 1, 2, \ldots$, constituting the search trajectory for starting point $\mathbf{x}^j$ will converge since $\Delta\mathbf{x}^i \to 0$ (see (4.70)). In the presence of many local minima, the probability of convergence to a relative low local minimum is increased, since the kinetic energy can only decrease for $\alpha < 1$.

Procedure *MP2*, for a given j, is successfully terminated if $||\nabla f(\mathbf{x}^i)|| \leq \epsilon$ for some small prescribed positive value ϵ, or when $\alpha V_0^i < \beta V_0^1$, and $\mathbf{x}^i$ is taken as the local minimizer $\mathbf{x}_f^j$ with corresponding function value $f_f^j := h(\mathbf{x}_f^j) - k$.

Clearly, the condition $\alpha V_0^i < \beta V_0^1$ will always occur for $0 < \beta < \alpha$ and $0 < \alpha < 1$.

MP2 can be viewed as a variant of the steepest descent algorithm. However, as opposed to steepest descent, MP2 has (as has MP1) the ability for 'hill-climbing', as is inherent in the physical model on which MP2 is based (viz., the trajectories of a bouncing ball in a conservative gravitational field.) Hence, the behavior of MP2 is quite different from that of steepest descent and furthermore, because of it's physical basis, it tends to seek local minima with relative low function values and is therefore suitable for implementation in global searches, while steepest descent is not.

For the MBB algorithm, convergence to a local minimum is not proven. Instead, the underlying physics of a bouncing ball is exploited. Unsuccessful trajectories are terminated, and do not contribute to the probabilistic stopping criterion (although these points are included in the number of unsuccessful trajectories $\tilde{n}$). In the validation of the algorithm the philosophy adopted here is that the practical demonstration of convergence of a proposed algorithm on a variety of demanding test problems may be as important and convincing as a rigorous mathematical convergence argument.

Indeed, although for the steepest descent method convergence can be proven, in practice it often fails to converge because effectively an infinite number of steps is required for convergence.

4.6.4 Global stopping criterion

The above methods require a termination rule for deciding when to end the sampling and to take the current overall minimum function value $\tilde{f}$, i.e.

$$\tilde{f} = \text{minimum}\left\{f_f^j,\ \text{over all } j \text{ to date}\right\} \tag{4.75}$$

as the global minimum value f^*.

Define the *region of convergence* of the dynamic methods for a local minimum $\hat{\mathbf{x}}$ as the set of all points $\mathbf{x}$ which, used as starting points for the above procedures, converge to $\hat{\mathbf{x}}$. One may reasonably expect that in the case where the *regions of attraction* (for the usual gradient-descent methods, see Dixon et al., 1976) of the local minima are more or less equal, that the region of convergence of the global minimum will be relatively increased.

Let R_k denote the region of convergence for the above minimization procedures *MP1* and *MP2* of local minimum $\hat{\mathbf{x}}^k$ and let α_k be the associated probability that a sample point be selected in R_k. The region of convergence and the associated probability for the global minimum $\mathbf{x}^*$ are denoted by R^* and α^* respectively. The following basic assumption, which is probably true for many functions of practical interest, is now made. BASIC ASSUMPTION:

$$\alpha^* \geq \alpha_k \text{ for all local minima } \hat{\mathbf{x}}^k. \tag{4.76}$$

The following theorem may be proved.

4.6.4.1 Theorem (Snyman and Fatti, 1987)

Let r be the number of sample points falling within the region of convergence of the current overall minimum $\tilde{f}$ after $\tilde{n}$ points have been sampled. Then under the above assumption and a statistically non-informative prior distribution the probability that $\tilde{f}$ corresponds to f^* may be obtained from

$$Pr\left[\tilde{f} = f^*\right] \geq q(\tilde{n}, r) = 1 - \frac{(\tilde{n}+1)!(2\tilde{n}-r)!}{(2\tilde{n}+1)!(\tilde{n}-r)!}. \tag{4.77}$$

On the basis of this theorem the *stopping rule* becomes: STOP when $Pr\left[\tilde{f} = f^*\right] \geq q^*$, where q^* is some prescribed desired confidence level, typically chosen as 0.99.

Proof:

We present here an outline of the proof of (4.77), and follow closely the presentation in Snyman and Fatti (1987). (We have since learned that the proof can be shown to be a generalization of the procedure proposed by Zielinsky, 1981.) Given $\tilde{n}^*$ and α^*, the probability that at least one point, $\tilde{n} \geq 1$, has converged to f^* is

$$\Pr[\tilde{n}^* \geq 1|\tilde{n}, r] = 1 - (1 - \alpha^*)^{\tilde{n}} . \tag{4.78}$$

In the Bayesian approach, we characterize our uncertainty about the value of α^* by specifying a prior probability distribution for it. This distribution is modified using the sample information (namely, $\tilde{n}$ and r) to form a posterior probability distribution. Let $p_*(\alpha^*|\tilde{n}, r)$ be the posterior probability distribution of α^*. Then,

$$\begin{aligned} \Pr[\tilde{n}^* \geq 1|\tilde{n}, r] &= \int_0^1 \left[1 - (1 - \alpha^*)^{\tilde{n}}\right] p_*(\alpha^*|\tilde{n}, r) d\alpha^* \\ &= 1 - \int_0^1 (1 - \alpha^*)^{\tilde{n}} p_*(\alpha^*|\tilde{n}, r) d\alpha^*. \end{aligned} \tag{4.79}$$

Now, although the r sample points converge to the current overall minimum, we do not know whether this minimum corresponds to the global minimum of f^*. Utilizing (4.76), and noting that $(1-\alpha)^{\tilde{n}}$ is a decreasing function of α, the replacement of α^* in the above integral by α yields

$$\Pr[\tilde{n}^* \geq 1|\tilde{n}, r] \geq \int_0^1 \left[1 - (1 - \alpha)^{\tilde{n}}\right] p(\alpha|\tilde{n}, r) d\alpha . \tag{4.80}$$

Now, using Bayes theorem we obtain

$$p(\alpha|\tilde{n}, r) = \frac{p(r|\alpha, \tilde{n}) p(\alpha)}{\int_0^1 p(r|\alpha, \tilde{n}) p(\alpha) d\alpha} . \tag{4.81}$$

Since the $\tilde{n}$ points are sampled at random and each point has a probability α of converging to the current overall minimum, r has a binomial distribution with parameters α and $\tilde{n}$. Therefore

$$p(r|\alpha, \tilde{n}) = \binom{\tilde{n}}{r} \alpha^r (1 - \alpha)^{\tilde{n} - r} . \tag{4.82}$$

Substituting (4.82) and (4.81) into (4.80) gives:

$$\Pr[\tilde{n}^* \geq 1|\tilde{n}, r] \geq 1 - \frac{\int_0^1 \alpha^r (1 - \alpha)^{2\tilde{n} - r} p(\alpha) d\alpha}{\int_0^1 \alpha^r (1 - \alpha)^{\tilde{n} - r} p(\alpha) d\alpha} . \tag{4.83}$$

A suitable flexible prior distribution $p(\alpha)$ for α is the beta distribution with parameters a and b. Hence,

$$p(\alpha) = [1/\beta(a,b)]\,\alpha^{a-1}(1-\alpha)^{b-1}, \qquad 0 \le \alpha \le 1. \tag{4.84}$$

Using this prior distribution gives:

$$\begin{aligned}
\Pr[\tilde{n}^* \ge 1|\tilde{n},r] &\ge 1 - \frac{\Gamma(\tilde{n}+a+b)\,\Gamma(2\tilde{n}-r+b)}{\Gamma(2\tilde{n}+a+b)\,\Gamma(\tilde{n}-r+b)} \\
&= 1 - \frac{(\tilde{n}+a+b-1)!\,(2\tilde{n}-r+b-1)!}{(2\tilde{n}+a+b-1)!\,(\tilde{n}-r+b-1)!}.
\end{aligned}$$

Assuming a prior expectation of 1, (viz. $a = b = 1$), we obtain

$$\Pr[\tilde{n}^* \ge 1|\tilde{n},r] = 1 - \frac{(\tilde{n}+1)!\,(2\tilde{n}-r)!}{(2\tilde{n}+1)!\,(\tilde{n}-r)!},$$

which is the required result. □

4.6.5 Numerical results

No.	Name	ID	n	Ref.
1	Griewank G1	G1	2	Törn and Zilinskas; Griewank
2	Griewank G2	G2	10	Törn and Zilinskas; Griewank
3	Goldstein-Price	GP	2	Törn and Zilinskas; Dixon and Szegö
4	Six-hump Camelback	C6	2	Törn and Zilinskas; Branin
5	Shubert, Levi No. 4	SH	2	Lucidi and Piccioni
6	Branin	BR	2	Törn and Zilinskas; Branin and Hoo
7	Rastrigin	RA	2	Törn and Zilinskas
8	Hartman 3	H3	3	Törn and Zilinskas; Dixon and Szegö
9	Hartman 6	H6	6	Törn and Zilinskas; Dixon and Szegö
10	Shekel 5	S5	4	Törn and Zilinskas; Dixon and Szegö
11	Shekel 7	S7	4	Törn and Zilinskas; Dixon and Szegö
12	Shekel 10	S10	4	Törn and Zilinskas; Dixon and Szegö

Table 4.6: The test functions

The test functions used are tabulated in Table 4.6, and tabulated numerical results are presented in Tables 4.7 and 4.8. In the tables, the reported number of function values N_f are the average of 10 independent (random) starts of each algorithm.

No.	ID	SF - This Study			SF - Previous		MBB		
		N_f	$(r/\tilde{n})_b$	$(r/\tilde{n})_w$	N_f	$r/\tilde{n}$	N_f	$(r/\tilde{n})_b$	$(r/\tilde{n})_w$
1	G1	4199	6/40	6/75	1606	6/20	2629	5/8	6/23
2	G2	25969	6/84	6/312	26076	6/60	19817	6/24	6/69
3	GP	2092	4/4	5/12	668	4/4	592	4/4	5/10
4	C6	426	4/4	5/9	263	4/4	213	4/4	5/10
5	SH	8491	6/29	6/104	—	—	1057	5/7	6/26
6	BR	3922	4/4	5/12	—	—	286	4/4	5/6
7	RA	4799	6/67	6/117	—	—	1873	4/4	6/42
8	H3	933	4/4	5/8	563	5/6	973	5/9	6/29
9	H6	1025	4/4	5/10	871	5/8	499	4/4	5/9
10	S5	1009	4/4	6/24	1236	6/17	2114	5/8	6/39
11	S7	1057	5/8	6/37	1210	6/17	2129	6/16	6/47
12	S10	845	4/4	6/31	1365	6/20	1623	5/7	6/39

Table 4.7: Numerical results

Method	Test Function					
	BR	C6	GP	RA	SH	H3
TRUST	55	31	103	59	72	58
MBB	25	29	74	168	171	24

Table 4.8: Cost (N_f) using *a priori* stopping condition

Unless otherwise stated, the following settings were used in the SF algorithm (see Snyman and Fatti, 1987): $\gamma = 2.0$, $\alpha = 0.95$, $\epsilon = 10^{-2}$, $\omega = 10^{-2}$, $\delta = 0.0$, $q^* = 0.99$, and $\Delta t = 1.0$. For the MBB algorithm, $\alpha = 0.99$, $\epsilon = 10^{-4}$, and $q^* = 0.99$ were used. For each problem, the initial velocity V_0 was chosen such that $\Delta \mathbf{x}^1$ was equal to half the 'radius' of the domain D. A local search strategy was implemented with varying α in the vicinity of local minima.

In Table 4.7, $(r/\tilde{n})_b$ and $(r/\tilde{n})_w$ respectively indicate the best and worst $r/\tilde{n}$ ratios (see equation (4.77)), observed during 10 independent optimization runs of both algorithms. The SF results compare well with the previously published results by Snyman and Fatti, who reported values for a single run only. For the Shubert, Branin and Rastrigin functions, the MBB algorithm is superior to the SF algorithm. For the Shekel functions (S5, S7 and S10), the SF algorithm is superior. As a result of the stopping criterion (4.77), the SF and MBB algorithms found the global optimum between 4 and 6 times for each problem.

The results for the trying Griewank functions (Table 4.7) are encouraging. G1 has some 500 local minima in the region of interest, and G2 several thousand. The values used for the parameters are as specified, with $\Delta t = 5.0$ for G1 and G2 in the SF-algorithm. It appears that both the SF and MBB algorithms are highly effective for problems with a large number of local minima in D, and problems with a large number of design variables.

In Table 4.8 the MBB algorithm is compared with the recently published deterministic TRUST algorithm (Barhen et al., 1997). Since the TRUST algorithm was terminated when the global approximation was within a specified tolerance of the (known) global optimum, a similar criterion was used for the MBB algorithm. The table reveals that the two algorithms compare well. Note however that the highest dimension of the test problems used in Barhen et al. (1997) is 3. It is unclear if the deterministic TRUST algorithm will perform well for problems of large dimension, or problems with a large number of local minima in D.

In conclusion, the numerical results indicate that both the Snyman-Fatti trajectory method and the modified bouncing ball trajectory method are effective in finding the global optimum efficiently. In particular, the results for the trying Griewank functions are encouraging. Both algorithms appear effective for problems with a large number of local minima in the domain, and problems with a large number of design variables. A salient feature of the algorithms is the availability of an apparently effective global stopping criterion.

Chapter 5

EXAMPLE PROBLEMS

5.1 Introductory examples

Problem 5.1.1

Sketch the geometrical solution to the optimization problem:

$$\begin{aligned}&\text{minimize } f(\mathbf{x}) = 2x_2 - x_1\\ &\text{subject to } g_1(\mathbf{x}) = x_1^2 + 4x_2^2 - 16 \leq 0,\\ &g_2(\mathbf{x}) = (x_1 - 3)^2 + (x_2 - 3)^2 - 9 \leq 0\\ &\text{and } x_1 \geq 0 \text{ and } x_2 \geq 0.\end{aligned}$$

In particular sketch the *contours of the objective function* and the *constraint curves.* Indicate the feasible region and the position of the *optimum* $\mathbf{x}^*$ and the active constraint(s).

The solution to this problem is indicated in Figure 5.1.

Problem 5.1.2

Consider the function $f(\mathbf{x}) = 100(x_2 - x_1^2)^2 + (1 - x_1)^2$.

(i) Compute the gradient vector and the Hessian matrix.

(ii) Show that $\nabla f(\mathbf{x}) = \mathbf{0}$ if $\mathbf{x}^* = [1, 1]^T$ and that $\mathbf{x}^*$ is indeed a strong local minimum.

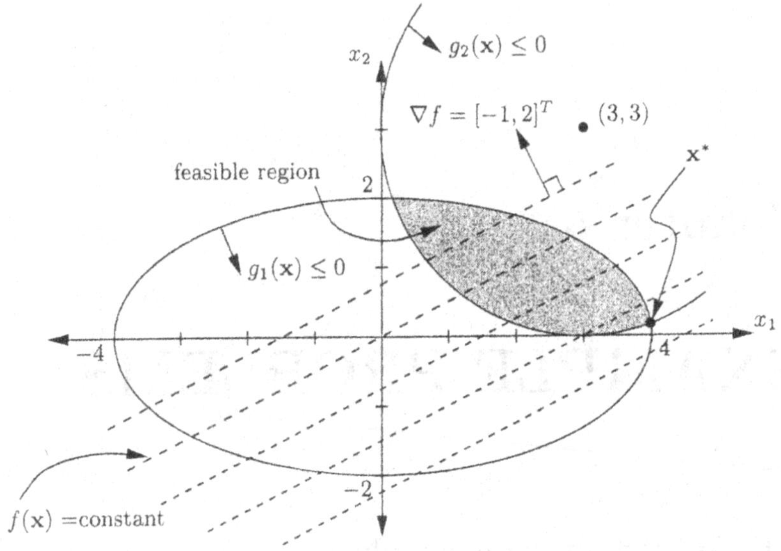

Figure 5.1: Solution to problem 5.1.1

(iii) Is $f(\mathbf{x})$ a convex function? Justify your answer.

Solution

(i)

$$\nabla f(\mathbf{x}) = \begin{bmatrix} 200(x_2 - x_1^2)(-2x_1) - 2(1 - x_1) \\ 200(x_2 - x_1^2) \end{bmatrix} = \begin{bmatrix} -400x_2x_1 + 400x_1^3 - 2 + 2x_1 \\ 200x_2 - 200x_1^2 \end{bmatrix}$$

$$\mathbf{H}(\mathbf{x}) = \begin{bmatrix} -400x_2 + 1200x_1^2 + 2 & \vdots & -400x_1 \\ -400x_1 & \vdots & 200 \end{bmatrix}.$$

(ii) $\nabla f(\mathbf{x}) = \mathbf{0}$ implies that

$$-400x_2x_1 + 400x_1^3 - 2 + 2x_1 = 0 \quad (5.1)$$
$$200x_2 - 200x_1^2 = 0. \quad (5.2)$$

From (5.2) $x_2 = x_1^2$, and substituting into (5.1) yields

$$-400x_1^3 + 400x_1^3 - 2 - 2x_1 = 0$$

giving $x_1 = 1$ and $x_2 = 1$ which is the unique solution.
Therefore at $\mathbf{x}^* = [1, 1]^T$:

$$\mathbf{H}(\mathbf{x}^*) = \begin{bmatrix} 802 & -400 \\ -400 & 200 \end{bmatrix},$$

and since the principal minors $\alpha_1 = 802 > 0$, and $\alpha_2 = \det \mathbf{H}(\mathbf{x}^*) = 400 > 0$, it follows by Sylvester's theorem that $\mathbf{H}(\mathbf{x})$ is positive-definite at $\mathbf{x}^*$ (see different equivalent definitions for positive definiteness that can be checked for, for example Fletcher (1987)). Thus $\mathbf{x}^* = [1, 1]^T$ is a strong local minimum since the necessary and sufficient conditions in (1.24) are both satisfied at $\mathbf{x}^*$.

(iii) By inspection, if $x_1 = 0$ and $x_2 = 1$ then $\mathbf{H} = \begin{bmatrix} -398 & 0 \\ 0 & 200 \end{bmatrix}$ and the determinant of the Hessian matrix is less than zero and $\mathbf{H}$ is not positive-definite at this point. Since by Theorem 6.1.2, $f(\mathbf{x})$ is convex over a set X *if and only if* $\mathbf{H}(\mathbf{x})$ is positive semi-definite for all $\mathbf{x}$ in X, it follows that $f(\mathbf{x})$ is not convex in this case.

Problem 5.1.3

Determine whether or not the following function is convex:

$$f(\mathbf{x}) = 4x_1^2 + 3x_2^2 + 5x_3^2 + 6x_1x_2 + x_1x_3 - 3x_1 - 2x_2 + 15.$$

Solution

The function $f(\mathbf{x})$ is convex if and only if the Hessian matrix $\mathbf{H}(\mathbf{x})$ is positive semi-definite at every point $\mathbf{x}$.

$$\text{Here } \nabla f(\mathbf{x}) = \begin{bmatrix} 8x_1 + 6x_2 - 3 + x_3 \\ 6x_2 + 6x_1 - 2 \\ 10x_3 + x_1 \end{bmatrix} \text{ and } \mathbf{H}(\mathbf{x}) = \begin{bmatrix} 8 & 6 & 1 \\ 6 & 6 & 0 \\ 1 & 0 & 10 \end{bmatrix}.$$

The principal minors are:

$$\alpha_1 = 8 > 0,\ \alpha_2 = \begin{vmatrix} 8 & 6 \\ 6 & 6 \end{vmatrix} = 12 > 0 \text{ and } \alpha_3 = |\mathbf{H}| = 114 > 0.$$

Thus by Sylvester's theorem $\mathbf{H}(\mathbf{x})$ is positive-definite and thus $f(\mathbf{x})$ is convex.

Problem 5.1.4

Determine all the stationary points of

$$f(\mathbf{x}) = x_1^3 + 3x_1x_2^2 - 3x_1^2 - 3x_2^2 + 4.$$

Classify each point according to whether it corresponds to maximum, minimum or saddle point.

Solution

The first order necessary condition for a stationary point is that

$$\nabla f = \begin{bmatrix} 3x_1^2 + 3x_2^2 - 6x_1 \\ 6x_1x_2 - 6x_2 \end{bmatrix} = \mathbf{0}$$

from which it follows that $6x_2(x_1 - 1) = 0$.

Therefore either $x_2 = 0$ or $x_1 = 1$ which respectively give:

$$\begin{array}{ccc} 3x_1^2 - 6x_1 = 0 & & 3 + 3x_2^2 - 6 = 0 \\ 3x_1(x_1 - 2) = 0 & \text{or} & x_2^2 = 1 \\ x_1 = 0;\ x_1 = 2 & & x_2 = \pm 1. \end{array}$$

Therefore the stationary points are: $(0;0)$, $(2;0)$; $(1;1)$, $(1;-1)$.

The nature of these stationary points may be determined by substituting their coordinates into the Hessian matrix $\mathbf{H}(\mathbf{x}) = \begin{bmatrix} 6x_1 - 6 & 6x_2 \\ 6x_2 & 6x_1 - 6 \end{bmatrix}$ and applying Sylvester's theorem.

The results are listed below.

Point	Hessian	Minors	Nature of **H**	Type
$(0;0)$	$\begin{bmatrix} -6 & 0 \\ 0 & -6 \end{bmatrix}$	$\alpha_1 < 0$ $\alpha_2 > 0$	negative-definite	maximum
$(2;0)$	$\begin{bmatrix} 6 & 0 \\ 0 & 6 \end{bmatrix}$	$\alpha_1 > 0$ $\alpha_2 > 0$	positive-definite	minimum
$(1;1)$	$\begin{bmatrix} 0 & 6 \\ 6 & 0 \end{bmatrix}$	$\alpha_1 = 0$ $\alpha_2 < 0$	indefinite	saddle
$(1;-1)$	$\begin{bmatrix} 0 & -6 \\ -6 & 0 \end{bmatrix}$	$\alpha_1 = 0$ $\alpha_2 < 0$	indefinite	saddle

Problem 5.1.5

Characterize the stationary points of $f(\mathbf{x}) = x_1^3 + x_2^3 + 2x_1^2 + 4x_2^2 + 6$.

Solution

First determine gradient vector $\nabla f(\mathbf{x})$ and consider $\nabla f(\mathbf{x}) = \mathbf{0}$:

$$\begin{aligned} 3x_1^2 + 4x_1 &= x_1(3x_1 + 4) = 0 \\ 3x_2^2 + 8x_2 &= x_2(3x_2 + 8) = 0. \end{aligned}$$

The solutions are: $(0,0)$; $\left(0,-\frac{8}{3}\right)$; $\left(-\frac{4}{3},0\right)$; $\left(-\frac{4}{3},-\frac{8}{3}\right)$.

To determine the nature of the stationary points substitute their coordinates in the Hessian matrix:

$$\mathbf{H}(\mathbf{x}) = \begin{bmatrix} 6x_1 + 4 & 0 \\ 0 & 6x_2 + 8 \end{bmatrix}$$

and study the principal minors α_1 and α_2 for each point:

Point $\mathbf{x}$	α_1	α_2	Nature of $\mathbf{H}(\mathbf{x})$	Type	$f(\mathbf{x})$
$(0,0)$	4	32	positive-definite	minimum	6
$(0,-\frac{8}{3})$	4	-32	indefinite	saddle	15.48
$(-\frac{4}{3},0)$	-4	-32	indefinite	saddle	7.18
$(-\frac{4}{3},-\frac{8}{3})$	-4	32	negative-definite	maximum	16.66

Problem 5.1.6

Minimize $f(\mathbf{x}) = x_1 - x_2 + 2x_1^2 + 2x_1x_2 + x_2^2$ by means of the basic Newton method. Use initial estimate for the minimizer $\mathbf{x}^0 = [0,0]^T$.

Solution

$$\nabla f(\mathbf{x}) = \begin{bmatrix} 1+4x_1+2x_2 \\ -1+2x_1+2x_2 \end{bmatrix} = \mathbf{g}(\mathbf{x})$$

$$\mathbf{H}(\mathbf{x}) = \left\{\frac{\partial^2 f}{\partial x_i \partial x_j}\right\} = \begin{bmatrix} 4 & 2 \\ 2 & 2 \end{bmatrix}; \quad \mathbf{H}^{-1} = \begin{bmatrix} \frac{1}{2} & -\frac{1}{2} \\ -\frac{1}{2} & 1 \end{bmatrix}.$$

First Newton iteration:

$$\begin{aligned} \mathbf{x}^1 &= \mathbf{x}^0 - \mathbf{H}^{-1}\mathbf{g}(\mathbf{x}^0) \\ &= \begin{bmatrix} 0 \\ 0 \end{bmatrix} - \begin{bmatrix} \frac{1}{2} & -\frac{1}{2} \\ -\frac{1}{2} & 1 \end{bmatrix} \begin{bmatrix} 1 \\ -1 \end{bmatrix} = \begin{bmatrix} -1 \\ \frac{3}{2} \end{bmatrix} \end{aligned}$$

and

$$\mathbf{g}(\mathbf{x}^1) = \begin{bmatrix} 1+4(-1)+2\left(\frac{3}{2}\right) \\ -1+2(-1)+2\left(\frac{3}{2}\right) \end{bmatrix} = \begin{bmatrix} 0 \\ 0 \end{bmatrix}.$$

Therefore since $\mathbf{H}$ is positive-definite for all $\mathbf{x}$, the sufficient conditions (1.24) for a strong local minimum at $\mathbf{x}^1$ are satisfied and the global minimum is $\mathbf{x}^* = \mathbf{x}^1$.

Problem 5.1.7

Minimize $f(\mathbf{x}) = 3x_1^2 - 2x_1x_2 + x_2^2 + x_1$ by means of the basic Newton method using $\mathbf{x}^0 = [1,1]^T$.

Solution

$$\mathbf{g}(\mathbf{x}) = \begin{bmatrix} \frac{\partial f}{\partial x_1} \\ \\ \frac{\partial f}{\partial x_2} \end{bmatrix} = \begin{bmatrix} 6x_1 - 2x_2 + 1 \\ -2x_1 + 2x_2 \end{bmatrix}$$

$$\mathbf{H}(\mathbf{x}) = \begin{bmatrix} 6 & -2 \\ -2 & 2 \end{bmatrix}; \text{ and } \mathbf{H}^{-1} = \begin{bmatrix} \frac{1}{4} & \frac{1}{4} \\ \\ \frac{1}{4} & \frac{3}{4} \end{bmatrix}.$$

First Newton iteration:

$$\begin{aligned} \mathbf{x}^1 &= \begin{bmatrix} 1 \\ 1 \end{bmatrix} - \mathbf{H}^{-1} \begin{bmatrix} 6-2+1 \\ -2+2 \end{bmatrix} \\ &= \begin{bmatrix} 1 \\ 1 \end{bmatrix} - \begin{bmatrix} \frac{1}{4} & \frac{1}{4} \\ \\ \frac{1}{4} & \frac{3}{4} \end{bmatrix} \begin{bmatrix} 5 \\ 0 \end{bmatrix} = \begin{bmatrix} 1-\frac{5}{4} \\ \\ 1-\frac{5}{4} \end{bmatrix} = \begin{bmatrix} -\frac{1}{4} \\ \\ -\frac{1}{4} \end{bmatrix}. \end{aligned}$$

With $\nabla f(\mathbf{x}^1) = \mathbf{0}$ and $\mathbf{H}$ positive-definite the necessary and sufficient conditions (1.24) are satisfied at $\mathbf{x}^1$ and therefore the global minimum is given by

$$\mathbf{x}^* = \left[-\frac{1}{4}, -\frac{1}{4}\right]^T \text{ and } f(\mathbf{x}^*) = -0.125.$$

5.2 Line search descent methods

Problem 5.2.1

Minimize $F(\lambda) = -\lambda \cos \lambda$ over the interval $\left[0, \frac{\pi}{2}\right]$ by means of the golden section method.

Solution

The golden ratio is $r = 0.618$ and $F(0) = F(\frac{\pi}{2}) = 0$.

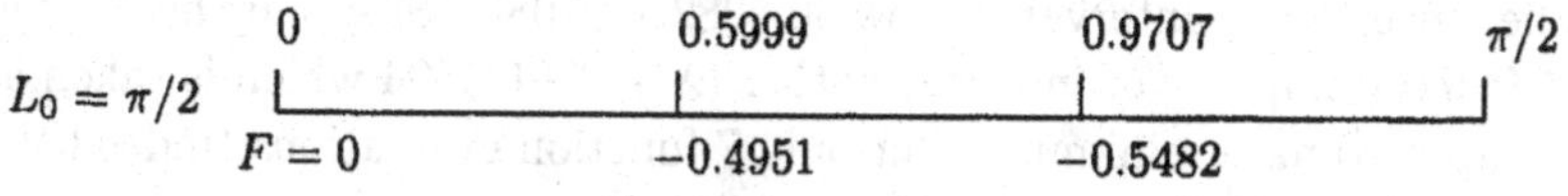

Figure 5.2

The first two interior points are: $rL_0 = 0.9707$ and $r^2L_0 = 0.5999$ with function values: $F(0.9707) = -0.5482$ and $F(0.5999) = -0.4951$.

Next new point$= 0.5999 + rL_1 = 1.200$ with $F(1.200) = -0.4350$.

	0.5999	0.9707	1.200	$\pi/2$
$L_1 = 0.9707$	$F = -0.4951$	-0.5482	-0.4350	

Figure 5.3

Next new point$= 0.5999 + r^2L_2 = 0.8292$ with $F(0.8292) = -0.5601$.

	0.5999	0.8292	0.9708	1.200
$L_2 = 0.6000$	$F = -0.4951$	-0.5601	-0.5482	-0.4350

Figure 5.4

Next new point$= 0.5999 + r^2L_3 = 0.7416$ with $F(0.7416) = -0.5468$.

	0.6	0.7416	0.8292	0.9708
$L_3 = 0.3708$	$F = -0.4951$	-0.5468	-0.5601	-0.5482

Figure 5.5

Next new point$= 0.7416 + rL_4 = 0.8832$ with $F(0.8832) = -0.5606$.

	0.7416	0.8292	0.8832	0.9708
$L_4 = 0.2292$	$F = -0.5468$	-0.5601	-0.5606	-0.5482

Figure 5.6

The uncertainty interval is now $[0.8282; 0.9708]$. Stopping here gives $\lambda^* \cong 0.9$ (midpoint of interval) with $F(\lambda^*) \cong -0.5594$ which is taken as the approximate minimum after only 7 function evaluations (indeed the actual $\lambda^* = 0.8603$ gives $F(\lambda^*) = -0.5611$).

Problem 5.2.2 Minimize the function $F(\lambda) = (\lambda - 1)(\lambda + 1)^2$, where λ is a single real variable, by means of Powell's quadratic interpolation method. Choose $\lambda_0 = 0$ and use $h = 0.1$. Perform *only two iterations.*

Solution

Set up difference table:

λ	$F(\lambda)$	$F[\,,\,]$	$F[\,,\,,\,]$
0	-1		
		-0.890	
0.1	-1.089		1.300
		-0.630	
0.2	-1.152		1.694
		-0.135	
$\lambda_m^{(1)} = 0.392$	-1.178		
$\lambda_m^{(2)} = 0.336$			

Turning point λ_m given by $\lambda_m = \dfrac{F[\,,\,,\,](\lambda_0 + \lambda_1) - F[\,,\,]}{2F[\,,\,,\,]}$:

First iteration:

$$\lambda_m^{(1)} = \frac{1.3(0.1) + 0.89}{2(1.3)} = 0.392$$

Second iteration:

$$\lambda_m^{(2)} = \frac{1.694(0.3) + 0.63}{2(1.694)} = 0.336.$$

Problem 5.2.3 Apply two steps of the *steepest descent method* to the minimization of $f(\mathbf{x}) = x_1 - x_2 + 2x_1^2 + 2x_1x_2 + x_2^2$. Use as starting point $\mathbf{x}^0 = [0, 0]^T$.

Solution

$$\nabla f(\mathbf{x}) = \begin{bmatrix} 1 + 4x_1 + 2x_2 \\ -1 + 2x_1 + 2x_2 \end{bmatrix}.$$

Step 1

$\mathbf{x}^0 = \begin{bmatrix} 0 \\ 0 \end{bmatrix}$ and the first steepest descent direction $\mathbf{u}^1 = -\nabla f(\mathbf{x}^0) = \begin{bmatrix} -1 \\ 1 \end{bmatrix}$.

Here it is convenient *not* to normalize.

The minimizer along the direction $\mathbf{u}^1$ is given by

$$\mathbf{x}^1 = \mathbf{x}^0 + \lambda_1 \mathbf{u}^1 = \begin{bmatrix} -\lambda_1 \\ \lambda_1 \end{bmatrix}.$$

To find λ_1, minimize with respect to λ the one variable function:

$$F(\lambda) = f(\mathbf{x}^0 + \lambda \mathbf{u}^1) = -\lambda - \lambda + 2\lambda^2 - 2\lambda^2 + \lambda^2.$$

The necessary condition for a minimum is $\frac{dF}{d\lambda} = 2\lambda - 2 = 0$ giving $\lambda_1 = 1$ and with $\frac{d^2F(\lambda_1)}{d\lambda^2} = 2 > 0$, λ_1 indeed corresponds to the minimum of $F(\lambda)$, and $\mathbf{x}^1 = \begin{bmatrix} -1 \\ 1 \end{bmatrix}$.

Step 2

The next steepest descent direction is

$$\mathbf{u}^2 = -\nabla f(\mathbf{x}^1) = \begin{bmatrix} 1 \\ 1 \end{bmatrix}$$

and

$$\mathbf{x}^2 = \mathbf{x}^1 + \lambda_2 \mathbf{u}^2 = \begin{bmatrix} -1 + \lambda_2 \\ 1 + \lambda_2 \end{bmatrix}.$$

To minimize in the direction of $\mathbf{u}^2$ consider

$$F(\lambda) = f(\mathbf{x}^1 + \lambda \mathbf{u}^2) = (-1+\lambda) - (1+\lambda) + 2(\lambda-1)^2 + 2(\lambda-1)(\lambda+1) + (\lambda+1)^2$$

and apply the necessary condition

$$\frac{dF}{d\lambda} = 10\lambda - 2 = 0.$$

This gives $\lambda_2 = \frac{1}{5}$ and with $\frac{d^2F(\lambda_2)}{d^2\lambda} = 10 > 0$ (minimum) it follows that

$$\mathbf{x}^2 = \begin{bmatrix} -1 + \frac{1}{5} \\ \\ 1 + \frac{1}{5} \end{bmatrix} = \begin{bmatrix} -0.8 \\ 1.2 \end{bmatrix}.$$

Problem 5.2.4

Apply two steps of the *steepest descent method* to the minimization of $f(\mathbf{x}) = (2x_1 - x_2)^2 + (x_2 + 1)^2$ with $\mathbf{x}^0 = [\frac{5}{2}, 2]^T$.

Solution

Step 1

$$\begin{aligned} \frac{\partial f}{\partial x_1} &= 4(2x_1 - x_2) \\ \frac{\partial f}{\partial x_2} &= 4x_2 - 4x_1 + 2 \end{aligned}$$

giving $\mathbf{g}^0 = \begin{bmatrix} 4(5-2) \\ 8 - 4(\frac{5}{2}) + 2 \end{bmatrix} = \begin{bmatrix} 12 \\ 0 \end{bmatrix}$.

After normalizing, the first search direction is $\mathbf{u}^1 = \begin{bmatrix} -1 \\ 0 \end{bmatrix}$ and $\mathbf{x}^1 = \begin{bmatrix} \frac{5}{2} - \lambda_1 \\ 2 \end{bmatrix}$. Now minimize with respect to λ:

$$F(\lambda) = f(\mathbf{x}^0 + \lambda\mathbf{u}^1) = \left(2\left(\tfrac{5}{2} - \lambda\right) - 2\right)^2 + (2+1)^2 = (3 - 2\lambda)^2 + 9.$$

The necessary condition $\frac{dF}{d\lambda} = 0$ gives $\lambda_1 = \frac{3}{2}$, and with $\frac{d^2F}{d\lambda^2}(\lambda_1) > 0$ ensuring a minimum, it follows that

$$\mathbf{x}^1 = \begin{bmatrix} \frac{5}{2} - \frac{3}{2} \\ 2 \end{bmatrix} \text{ and } \mathbf{g}^1 = \begin{bmatrix} 4(2-2) \\ 8 - 4 + 2 \end{bmatrix} = \begin{bmatrix} 0 \\ 6 \end{bmatrix}.$$

Step 2

New normalized steepest descent direction: $\mathbf{u}^2 = \begin{bmatrix} 0 \\ -1 \end{bmatrix}$, thus $\mathbf{x}^2 = \begin{bmatrix} 1 \\ 2 - \lambda_2 \end{bmatrix}$.

In the direction of $\mathbf{u}^2$:

$$F(\lambda) = f(\mathbf{x}^1 + \lambda\mathbf{u}^2) = \lambda^2 + (3 - \lambda)^2.$$

Setting $\frac{dF}{d\lambda} = 2\lambda - 2(3 - \lambda) = 0$ gives $\lambda_2 = \frac{3}{2}$ and since $\frac{d^2F}{d\lambda^2}(\lambda_2) > 0$ a minimum is ensured, and therefore

$$\mathbf{x}^2 = \begin{bmatrix} 1 \\ 2 - \frac{3}{2} \end{bmatrix} = \begin{bmatrix} 1 \\ \frac{1}{2} \end{bmatrix}.$$

Problem 5.2.5

Apply the *Fletcher-Reeves method* to the minimization of

$$f(\mathbf{x}) = (2x_1 - x_2)^2 + (x_2 + 1)^2 \text{ with } \mathbf{x}^0 = \left[\tfrac{5}{2}, 2\right]^T.$$

Solution

The first step is identical to that of the steepest descent method given for Problem 5.2.4 above, with $\mathbf{u}^1 = -\mathbf{g}^0 = [-12, 0]^T$.

For the *second step* the *Fletcher-Reeves* search direction becomes

$$\mathbf{u}^2 = -\mathbf{g}^1 + \frac{\|\mathbf{g}^1\|^2}{\|\mathbf{g}^0\|^2}\mathbf{u}^1.$$

Using the data from the first step in Problem 5.2.4, the second search direction becomes

$$\mathbf{u}^2 = \begin{bmatrix} 0 \\ -6 \end{bmatrix} + \tfrac{36}{144}\begin{bmatrix} -12 \\ 0 \end{bmatrix} = \begin{bmatrix} -3 \\ -6 \end{bmatrix}$$

and

$$\mathbf{x}^2 = \mathbf{x}^1 + \lambda_2\mathbf{u}^2 = \begin{bmatrix} 1 \\ 2 \end{bmatrix} - \begin{bmatrix} 3\lambda_2 \\ 6\lambda_2 \end{bmatrix} = \begin{bmatrix} 1 - 3\lambda_2 \\ 2 - 6\lambda_2 \end{bmatrix}.$$

In the direction $\mathbf{u}^2$:

$$F(\lambda) = (2(1 - 3\lambda) - 2 + 6\lambda)^2 + (2 - 6\lambda + 1)^2 = (3 - 6\lambda)^2$$

and the necessary condition for a minimum is

$$\frac{dF}{d\lambda} = -12(3 - 6\lambda) = 0 \text{ giving } \lambda_2 = \tfrac{1}{2}$$

and thus with $\frac{d^2F(\lambda_2)}{d\lambda^2} = 36 > 0$:

$$\mathbf{x}^2 = \begin{bmatrix} 1 - \frac{3}{2} \\ \\ 2 - \frac{6}{2} \end{bmatrix} = \begin{bmatrix} -\frac{1}{2} \\ -1 \end{bmatrix}$$

with

$$\mathbf{g}^2 = \begin{bmatrix} 4\left(2\left(-\frac{1}{2}\right) - 1(-1)\right) \\ 4(-1) - 4\left(-\frac{1}{2}\right) + 2 \end{bmatrix} = \begin{bmatrix} 0 \\ 0 \end{bmatrix}.$$

Since $\mathbf{g}^2 = \mathbf{0}$ and $\mathbf{H} = \begin{bmatrix} 8 & -4 \\ -4 & 4 \end{bmatrix}$ is positive-definite, $\mathbf{x}^2 = \mathbf{x}^*$.

Problem 5.2.6

Minimize $F(\mathbf{x}) = x_1 - x_2 + 2x_1^2 + 2x_1x_2 + x_2^2$ by using the *Fletcher-Reeves* method. Use as starting point $\mathbf{x}^0 = [0, 0]^T$.

Solution

Step 1

$$\nabla f(\mathbf{x}) = \begin{bmatrix} 1 + 4x_1 + 2x_2 \\ -1 + 2x_1 + 2x_2 \end{bmatrix}, \quad \mathbf{x}^0 = [0, 0]^T,$$

$$\mathbf{g}^0 = \begin{bmatrix} 1 \\ -1 \end{bmatrix}, \quad \mathbf{u}^1 = -\mathbf{g}^0 = \begin{bmatrix} -1 \\ 1 \end{bmatrix} \text{ and } \mathbf{x}^1 = \mathbf{x}^0 + \lambda_1\mathbf{u}^1 = \begin{bmatrix} -\lambda_1 \\ \lambda_1 \end{bmatrix}.$$

Therefore

$$F(\lambda) = f(\mathbf{x}^0 + \lambda\mathbf{u}^1) = -\lambda - \lambda + 2\lambda^2 - 2\lambda^2 + \lambda^2 = \lambda^2 - 2\lambda$$

and $\frac{dF}{d\lambda} = 2\lambda - 2 = 0$ giving $\lambda_1 = 1$ and with $\frac{d^2F(\lambda_1)}{d\lambda^2} = 2 > 0$ ensuring a minimum, it follows that $\mathbf{x}^1 = [-1, 1]^T$.

Step 2

$$\mathbf{g}^1 = \nabla f(\mathbf{x}^1) = \begin{bmatrix} -1 \\ -1 \end{bmatrix}$$

$$\mathbf{u}^2 = -\mathbf{g}^1 + \frac{\|\mathbf{g}^1\|^2}{\|\mathbf{g}^0\|^2}\mathbf{u}^1 = \begin{bmatrix} 1 \\ 1 \end{bmatrix} + \frac{2}{2}\begin{bmatrix} -1 \\ 1 \end{bmatrix} = \begin{bmatrix} 0 \\ 2 \end{bmatrix}$$

$$\mathbf{x}^2 = \begin{bmatrix} -1 \\ 1 \end{bmatrix} + \lambda_2\begin{bmatrix} 0 \\ 2 \end{bmatrix} = [-1, 1 + 2\lambda_2]^T.$$

Thus

$$F(\lambda) = -1 - (1 + 2\lambda) + 2(-1)^2 + 2(-1)(1 + 2\lambda) + (1 + 2\lambda)^2 = 4\lambda^2 - 2\lambda - 1$$

with $\frac{dF}{d\lambda} = 8\lambda - 2 = 0$ giving $\lambda_2 = \frac{1}{4}$ and therefore, with $\frac{d^2F}{d\lambda^2} = 8 > 0$:

$$\mathbf{x}^2 = \begin{bmatrix} -1 \\ 1 \end{bmatrix} + \frac{1}{4}\begin{bmatrix} 0 \\ 2 \end{bmatrix} = [-1, 1.5]^T.$$

This results in

$$\mathbf{g}^2 = \begin{bmatrix} 1 - 4 + 2(1.5) \\ -1 - 2 + 2(1.5) \end{bmatrix} = [0, 0]^T$$

and since $\mathbf{H}$ is positive-definite the optimum solution is $\mathbf{x}^* = \mathbf{x}^2 = [-1, 1.5]^T$.

Problem 5.2.7

Minimize $f(\mathbf{x}) = x_1^2 - x_1x_2 + 3x_2^2$ with $\mathbf{x}^0 = (1, 2)^T$ by means of the *Fletcher-Reeves* method.

Solution

Step 1

Since $\mathbf{g}^0 = \begin{bmatrix} 2x_1 - x_2 \\ 6x_2 - x_1 \end{bmatrix} = \begin{bmatrix} 0 \\ 11 \end{bmatrix}$, $\mathbf{u}^1 = -\mathbf{g}^0 = [0, -11]^T$ and $\mathbf{x}^1 = [1, 2]^T + \lambda_1[0, -11]^T$.

This results in

$$F(\lambda) = 1 - 1(2 - 11\lambda) + 3(2 - 11\lambda)^2$$

with

$$\frac{dF}{d\lambda} = 11 + 6(2 - 11\lambda)(-11) = 0 \text{ giving } \lambda_1 = \tfrac{1}{6}.$$

Thus, with $\frac{d^2F(\lambda_1)}{d\lambda^2} > 0$: $\mathbf{x}^1 = [1, 2]^T + \frac{1}{6}[0, -11]^T = \left[1, \frac{1}{6}\right]^T$.

Step 2

$\mathbf{g}^1 = \begin{bmatrix} \frac{11}{6} \\ 0 \end{bmatrix}$ and thus

$$\mathbf{u}^2 = -\mathbf{g}^1 + \frac{\|\mathbf{g}^1\|^2}{\|\mathbf{g}^0\|^2}\mathbf{u}^1 = \begin{bmatrix} -\frac{11}{6} \\ 0 \end{bmatrix} + \frac{1}{36}\begin{bmatrix} 0 \\ -11 \end{bmatrix} = \begin{bmatrix} -\frac{11}{6} \\ \\ -\frac{11}{36} \end{bmatrix}$$

giving

$$\mathbf{x}^2 = \mathbf{x}^1 + \lambda_2\mathbf{u}^2 = \begin{bmatrix} 1 \\ \frac{1}{6} \end{bmatrix} - \lambda_2 \begin{bmatrix} \frac{11}{6} \\ \\ \frac{11}{36} \end{bmatrix} = \begin{bmatrix} \left(1 - \lambda_2\frac{11}{6}\right) \\ \\ \frac{1}{6}\left(1 - \lambda_2\frac{11}{6}\right) \end{bmatrix}$$

Thus

$$\begin{aligned} F(\lambda) &= \left(1-\tfrac{11}{6}\lambda\right)^2 - \tfrac{1}{6}\left(1-\tfrac{11}{6}\lambda\right)^2 + \tfrac{3}{36}\left(1-\tfrac{11}{6}\lambda\right)^2 \\ &= \left(1-\tfrac{1}{6}+\tfrac{1}{12}\right)\left(1-\tfrac{11}{6}\lambda\right)^2 \end{aligned}$$

and $\frac{dF}{d\lambda} = \frac{11}{6}\left(1-\frac{11}{6}\lambda\right)\left(-\frac{11}{6}\right) = 0$ gives $\lambda_2 = \frac{6}{11}$, and with $\frac{d^2F(\lambda_2)}{d\lambda^2} > 0$ gives $\mathbf{x}^2 = [0,0]^T$. With $\mathbf{g}^2 = \mathbf{0}$, and since $\mathbf{H}$ is positive-definite for all $\mathbf{x}$, this is the optimal solution.

Problem 5.2.8

Obtain the first updated matrix $\mathbf{G}_1$ when applying the DFP method to the minimization of $f(\mathbf{x}) = 4x_1^2 - 40x_1 + x_2^2 - 12x_2 + 136$ with starting point
$\mathbf{x}^0 = [8,9]^T$.

Solution

Factorizing $f(\mathbf{x})$ gives $f(\mathbf{x}) = 4(x_1-5)^2 + (x_2-6)^2$ and $\nabla f(\mathbf{x}) = \begin{bmatrix} 8x_1 - 40 \\ 2x_2 - 12 \end{bmatrix}$.

Step 1

Choose $\mathbf{G}_0 = \mathbf{I}$, then for $\mathbf{x}^0 = [8,9]^T$, $\mathbf{g}^0 = [24,6]^T$.

$$\mathbf{u}^1 = -\mathbf{I}\mathbf{g}^0 = -\begin{bmatrix} 24 \\ 6 \end{bmatrix}$$

and $\mathbf{x}^1 = \mathbf{x}^0 + \lambda_2\mathbf{u}^1 = \begin{bmatrix} 8-24\lambda_2 \\ 9-6\lambda_2 \end{bmatrix}$.

The function to be minimized with respect to λ is

$$F(\lambda) = f(\mathbf{x}^0 + \lambda\mathbf{u}^1) = 4(8-24\lambda-5)^2 + (9-6\lambda-6)^2.$$

The necessary condition for a minimum, $\frac{dF}{d\lambda} = -8(24)(3-24\lambda)+2(-6)(3-6\lambda) = 0$ yields $\lambda_1 = 0.1308$ and thus with $\frac{d^2F(\lambda_1)}{d\lambda^2} > 0$:

$$\mathbf{x}^1 = \begin{bmatrix} 8-0.1308(24) \\ 9-0.1308(6) \end{bmatrix} = \begin{bmatrix} 4.862 \\ 8.215 \end{bmatrix} \text{ with } \mathbf{g}^1 = \nabla f(\mathbf{x}^1) = \begin{bmatrix} -1.10 \\ 4.43 \end{bmatrix}.$$

The DFP update now requires the following quantities:

$$\mathbf{v}^1 = \lambda_1 \mathbf{u}^1 = \mathbf{x}^1 - \mathbf{x}^0 = \begin{bmatrix} 4.862 - 8 \\ 8.215 - 9 \end{bmatrix} = \begin{bmatrix} -3.14 \\ -0.785 \end{bmatrix}$$

and $\mathbf{y}^1 = \mathbf{g}^1 - \mathbf{g}^0 = \begin{bmatrix} -25.10 \\ -1.57 \end{bmatrix}$

giving $\mathbf{v}^{1T}\mathbf{y}^1 = [-3.14, -0.785] \begin{bmatrix} -25.10 \\ -1.57 \end{bmatrix} = 80.05$

and $\mathbf{y}^{1T}\mathbf{y}^1 = [(25.10)^2 + (-1.57)^2] = 632.47$

to be substituted in the update formula (2.16):

$$\begin{aligned} \mathbf{G}_1 &= \mathbf{G}_0 + \frac{\mathbf{v}^1\mathbf{v}^{1T}}{\mathbf{v}^{1T}\mathbf{y}^1} - \frac{\mathbf{y}^1\mathbf{y}^{1T}}{\mathbf{y}^{1T}\mathbf{y}^1} \\ &= \begin{bmatrix} 1 & 0 \\ 0 & 1 \end{bmatrix} + \tfrac{1}{80.05} \begin{bmatrix} -3.14 \\ -0.785 \end{bmatrix} [-3.14; -0.785] \\ &\quad - \tfrac{1}{632.47} \begin{bmatrix} -25.10 \\ -1.57 \end{bmatrix} [-25.10; -1.57] \\ &= \begin{bmatrix} 1 & 0 \\ 0 & 1 \end{bmatrix} + \tfrac{1}{80.05} \begin{bmatrix} 9.860 & 2.465 \\ 2.465 & 0.6161 \end{bmatrix} - \tfrac{1}{632.47} \begin{bmatrix} 630.01 & 3941 \\ 39.41 & 2.465 \end{bmatrix} \\ &= \begin{bmatrix} 0.127 & -0.032 \\ -0.032 & 1.004 \end{bmatrix}. \end{aligned}$$

Problem 5.2.9

Determine the first updated matrix $\mathbf{G}_1$ when applying the DFP method to the minimization of $f(\mathbf{x}) = 3x_1^2 - 2x_1x_2 + x_2^2 + x_1$ with $\mathbf{x}^0 = [1, 1]^T$.

Solution

$$\frac{\partial f}{\partial x_1} = 6x_1 - 2x_2 + 1$$

$$\frac{\partial f}{\partial x_2} = -2x_1 + 2x_2.$$

Step 1

$\mathbf{g}^0 = [5, 0]^T$ and $\mathbf{G}_0 = \mathbf{I}$ which results in

$$\mathbf{x}^1 = \mathbf{x}^0 + \lambda_1 \mathbf{u}^1 = \mathbf{x}^0 - \lambda_1 \mathbf{G}_0 \mathbf{g}^0 = \begin{bmatrix} 1 \\ 1 \end{bmatrix} - \lambda_1 \begin{bmatrix} 1 & 0 \\ 0 & 1 \end{bmatrix} \begin{bmatrix} 5 \\ 0 \end{bmatrix} = \begin{bmatrix} 1 - 5\lambda_1 \\ 1 \end{bmatrix}.$$

For the function $F(\lambda) = 3(1-5\lambda)^2 - 2(1-5\lambda) + 1 + (1-5\lambda)$ the necessary condition for a minimum is $\frac{\partial F}{d\lambda} = 6(1-5\lambda)(-5) + 10 - 5 = 0$, which gives $\lambda_1 = \frac{25}{150} = \frac{1}{6}$, and since $\frac{d^2F(\lambda_1)}{d\lambda^2} > 0$:

$$\mathbf{x}^1 = \begin{bmatrix} 1 - \frac{5}{6} \\ 1 \end{bmatrix} = \begin{bmatrix} \frac{1}{6} \\ 1 \end{bmatrix}$$

and $\mathbf{v}^1 = \mathbf{x}^1 - \mathbf{x}^0 = \begin{bmatrix} \frac{1}{6} \\ 1 \end{bmatrix} - \begin{bmatrix} 1 \\ 1 \end{bmatrix} = \begin{bmatrix} -\frac{5}{6} \\ 0 \end{bmatrix}$,

with $\mathbf{y}^1 = \mathbf{g}^1 - \mathbf{g}^0 = \begin{bmatrix} 6\left(\frac{1}{6}\right) - 2 + 1 \\ -2\left(\frac{1}{6}\right) + 2 \end{bmatrix} - \begin{bmatrix} 5 \\ 0 \end{bmatrix} = \begin{bmatrix} -5 \\ \frac{5}{3} \end{bmatrix}$.

It follows that

$$\mathbf{v}^1\mathbf{v}^{1T} = \begin{bmatrix} -\frac{5}{6} \\ 0 \end{bmatrix} [-\tfrac{5}{6}, 0] = \begin{bmatrix} \frac{25}{36} & 0 \\ 0 & 0 \end{bmatrix}, \quad \mathbf{v}^{1T}\mathbf{y}^1 = [-\tfrac{5}{6}, 0] \begin{bmatrix} -5 \\ \frac{5}{3} \end{bmatrix} = \frac{25}{6},$$

$$\mathbf{y}^1\mathbf{y}^{1T} = \begin{bmatrix} -5 \\ \frac{5}{3} \end{bmatrix} [-5, \tfrac{5}{3}] = \begin{bmatrix} 25 & -\frac{25}{3} \\ -\frac{25}{3} & \frac{25}{9} \end{bmatrix}$$

and $\mathbf{y}^{1T}\mathbf{y}^1 = [-5, \frac{5}{3}] \begin{bmatrix} -5 \\ \frac{5}{3} \end{bmatrix} = 25 + \frac{25}{9} = \frac{250}{9}$.

In the above computations $\mathbf{G}_0$ has been taken as $\mathbf{G}_0 = \mathbf{I}$.

Substituting the above results in the update formula (2.16) yields $\mathbf{G}_1$ as follows:

$$\begin{aligned}
\mathbf{G}_1 &= \mathbf{G}_0 + \mathbf{A}_1 + \mathbf{B}_1 \\
&= \begin{bmatrix} 1 & 0 \\ 0 & 1 \end{bmatrix} + \frac{6}{25}\begin{bmatrix} \frac{25}{36} & 0 \\ 0 & 1 \end{bmatrix} - \frac{9}{250}\begin{bmatrix} 25 & -\frac{25}{3} \\ -\frac{25}{3} & \frac{25}{9} \end{bmatrix} \\
&= \begin{bmatrix} 1 & 0 \\ 0 & 1 \end{bmatrix} + \begin{bmatrix} \frac{1}{6} & 0 \\ 0 & 1 \end{bmatrix} - \frac{9}{10}\begin{bmatrix} 1 & -\frac{1}{3} \\ -\frac{1}{3} & \frac{1}{9} \end{bmatrix} \\
&= \begin{bmatrix} \frac{7}{9} - \frac{9}{10} & \frac{3}{10} \\ \frac{3}{10} & \frac{9}{10} \end{bmatrix} = \begin{bmatrix} \frac{4}{15} & \frac{3}{10} \\ \frac{3}{10} & \frac{9}{10} \end{bmatrix}.
\end{aligned}$$

Comparing $\mathbf{G}_1$ with $\mathbf{H}^{-1} = \begin{bmatrix} \frac{1}{4} & \frac{1}{4} \\ \frac{1}{4} & \frac{1}{4} \end{bmatrix}$ shows that after only one iteration a reasonable approximation to the inverse has already been obtained.

Problem 5.2.10

Apply the DFP-method to the minimization $f(\mathbf{x}) = x_1 - x_2 + 2x_1^2 + 2x_1x_2 + x_2^2$ with starting point $\mathbf{x}^0 = [0,0]^T$.

Solution

Step 1

$\mathbf{G}_0 = \mathbf{I} = \begin{bmatrix} 1 & 0 \\ 0 & 1 \end{bmatrix}$ and $\mathbf{g}^0 = \begin{bmatrix} 1 + 4x_1 + 2x_2 \\ -1 + 2x_1 + 2x_2 \end{bmatrix} = \begin{bmatrix} 1 \\ -1 \end{bmatrix}$ gives

$\mathbf{u}^1 = -\mathbf{G}_0\mathbf{g}^0 = -\mathbf{I}\begin{bmatrix} 1 \\ -1 \end{bmatrix} = \begin{bmatrix} -1 \\ 1 \end{bmatrix}$ and thus

$\mathbf{x}^1 = \mathbf{x}^0 + \lambda_1\mathbf{u}^1 = \begin{bmatrix} -\lambda_1 \\ \lambda_1 \end{bmatrix}$.

Thus $F(\lambda) = \lambda^2 - 2\lambda$, and $\frac{dF}{d\lambda} = 0$ yields $\lambda_1 = 1$, and with $\frac{d^2F(\lambda_1)}{d\lambda^2} > 0$:

$\mathbf{x}^1 = \begin{bmatrix} -1 \\ 1 \end{bmatrix}$ and $\mathbf{g}^1 = \begin{bmatrix} -1 \\ -1 \end{bmatrix}$.

Now, using $\mathbf{v}^1 = \begin{bmatrix} -1 \\ 1 \end{bmatrix}$ and $\mathbf{y}^1 = \mathbf{g}^1 - \mathbf{g}^0 = \begin{bmatrix} -2 \\ 0 \end{bmatrix}$

in the update formula (2.16), gives

$$\mathbf{A}_1 = \frac{\mathbf{v}^1\mathbf{v}^{1T}}{\mathbf{v}^{1T}\mathbf{y}^1} = \frac{\begin{bmatrix} -1 \\ 1 \end{bmatrix}[-1\ 1]}{2} = \tfrac{1}{2}\begin{bmatrix} 1 & -1 \\ -1 & 1 \end{bmatrix}, \text{ and since } \mathbf{G}_0\mathbf{y}^1 = \mathbf{y}^1:$$

$$\mathbf{B}_1 = -\frac{\mathbf{y}^1\mathbf{y}^{1T}}{\mathbf{y}^{1T}\mathbf{y}^1} = -\tfrac{1}{4}\begin{bmatrix} 4 & 0 \\ 0 & 0 \end{bmatrix}.$$

Substitute the above in (2.16):

$$\begin{aligned} \mathbf{G}_1 &= \mathbf{G}_0 + \mathbf{A}_1 + \mathbf{B}_1 = \begin{bmatrix} 1 & 0 \\ 0 & 1 \end{bmatrix} + \tfrac{1}{2}\begin{bmatrix} 1 & -1 \\ -1 & 1 \end{bmatrix} - \tfrac{1}{4}\begin{bmatrix} 4 & 0 \\ 0 & 0 \end{bmatrix} \\ &= \begin{bmatrix} \frac{1}{2} & -\frac{1}{2} \\ -\frac{1}{2} & \frac{3}{2} \end{bmatrix}. \end{aligned}$$

Step 2

New search direction $\mathbf{u}^2 = -\mathbf{G}_1\mathbf{g}^1 = -\begin{bmatrix} \frac{1}{2} & -\frac{1}{2} \\ -\frac{1}{2} & \frac{3}{2} \end{bmatrix}\begin{bmatrix} -1 \\ -1 \end{bmatrix} = \begin{bmatrix} 0 \\ 1 \end{bmatrix}$

and therefore

$$\mathbf{x}^2 = \begin{bmatrix} -1 \\ 1 \end{bmatrix} + \lambda_2\begin{bmatrix} 0 \\ 1 \end{bmatrix} = [-1, 1 + \lambda_2]^T.$$

Minimizing $F(\lambda) = -1 - (1 + \lambda) + 2 - 2(1 + \lambda) + (1 + \lambda)^2$ implies $\frac{dF}{d\lambda} = -1 - 2 + 2(1 + \lambda) = 0$ which gives $\lambda_2 = \frac{1}{2}$.

Thus, since $\frac{d^2F(\lambda_2)}{d\lambda^2} > 0$, the minimum is given by $\mathbf{x}^2 = \begin{bmatrix} -1 \\ 1 \end{bmatrix} + \frac{1}{2}\begin{bmatrix} 0 \\ 1 \end{bmatrix} = \begin{bmatrix} -1 \\ \frac{3}{2} \end{bmatrix}$ with $\mathbf{g}^2 = \begin{bmatrix} 0 \\ 0 \end{bmatrix}$ and therefore $\mathbf{x}^2$ is optimal.

5.3 Standard methods for constrained optimization

5.3.1 Penalty function problems

Problem 5.3.1.1

Determine the shortest distance from the origin to the plane $x_1 + x_2 + x_3 = 1$ by means of the penalty function method.

Solution

Notice that the problem is equivalent to the problem:

minimize $f(\mathbf{x}) = x_1^2 + x_2^2 + x_3^2$ such that $x_1 + x_2 + x_3 = 1$.

The appropriate penalty function is $P = x_1^2 + x_2^2 + x_3^2 + \rho(x_1 + x_2 + x_3 - 1)^2$.

The necessary conditions at an unconstrained minimum of P are

$$\begin{aligned}
\frac{\partial P}{\partial x_1} &= 2x_1 + 2\rho(x_1 + x_2 + x_3 - 1) = 0\\
\frac{\partial P}{\partial x_2} &= 2x_2 + 2\rho(x_1 + x_2 + x_3 - 1) = 0\\
\frac{\partial P}{\partial x_3} &= 2x_3 + 2\rho(x_1 + x_2 + x_3 - 1) = 0.
\end{aligned}$$

Clearly $x_1 = x_2 = x_3$, and it follows that $x_1 = -\rho(3x_1 - 1)$, i.e.

$x_1(\rho) = \frac{\rho}{1+3\rho} = \frac{1}{3+\frac{1}{\rho}}$ and $\lim_{\rho\to\infty} x_1(\rho) = \frac{1}{3}$.

The shortest distance is therefore$= \sqrt{x_1^2 + x_2^2 + x_3^2} = \sqrt{3(\frac{1}{3})^2} = \frac{1}{\sqrt{3}}$.

Problem 5.3.1.2

Apply the *penalty function method* to the problem:

minimize $f(\mathbf{x}) = (x_1 - 1)^2 + (x_2 - 2)^2$

such that

$$h(\mathbf{x}) = x_2 - x_1 - 1 = 0,\ g(\mathbf{x}) = x_1 + x_2 - 2 \leq 0,\ -x_1 \leq 0,\ -x_2 \leq 0.$$

Solution

The appropriate penalty function is

$$P = (x_1-1)^2+(x_2-2)^2+\rho(x_2-x_1-1)^2+\beta_1(x_1+x_2-2)^2+\beta_2 x_1^2+\beta_3 x_2^2$$

where $\rho \gg 0$ and $\beta_j = \rho$ if the corresponding inequality constraint is violated, otherwise $\beta_j = 0$.

Clearly the unconstrained minimum $[1,2]^T$ violates the constraint $g(\mathbf{x}) \leq 0$, therefore assume that $\mathbf{x}$ is in the first quadrant but outside the feasible region, i.e. $\beta_1 = \rho$. The penalty function then becomes

$$P = (x_1 - 1)^2 + (x_2 - 2)^2 + \rho(x_2 - x_1 - 1)^2 + \rho(x_1 + x_2 - 2)^2.$$

The necessary conditions at the unconstrained minimum of P are:

$$\begin{aligned}\frac{\partial P}{\partial x_1} &= 2(x_1 - 1) - 2\rho(x_2 - x_1 - 1) + 2\rho(x_2 + x_1 - 2) = 0\\ \frac{\partial P}{\partial x_2} &= 2(x_2 - 1) + 2\rho(x_2 - x_1 - 1) + 2\rho(x_2 + x_1 - 2) = 0.\end{aligned}$$

The first condition is $x_1(2+4\rho) - 2 - 2\rho = 0$, from which it follows that

$$x_1(\rho) = \frac{2\rho+2}{4\rho+2} = \frac{2+\frac{2}{\rho}}{4+\frac{2}{\rho}} \text{ and } \lim_{\rho\to\infty} x_1(\rho) = \tfrac{1}{2}.$$

The second condition is $x_2(2 + 4\rho) - 4 - 6\rho = 0$, which gives

$$x_2(\rho) = \frac{6\rho+4}{4\rho+2} = \frac{6+\frac{4}{\rho}}{4+\frac{2}{\rho}} \text{ and } \lim_{\rho\to\infty} x_2(\rho) = \tfrac{3}{2}.$$

The optimum is therefore $\mathbf{x}^* = [\frac{1}{2}, \frac{3}{2}]^T$.

Problem 5.3.1.3

Apply the *penalty function method* to the problem:

minimize $f(\mathbf{x}) = x_1^2 + 2x_2^2$ such that $g(\mathbf{x}) = 1 - x_1 - x_2 \leq 0$.

Solution

The penalty function is $P = x_1^2 + 2x_2^2 + \beta(1 - x_1 - x_2)^2$. Again the unconstrained minimum clearly violates the constraint, therefore assume the constraint is violated in the penalty function, i.e. $\beta = \rho$. The necessary conditions at an unconstrained minimum of P are

$$\begin{aligned}\frac{\partial P}{\partial x_1} &= 2x_1 - 2\rho(1 - x_1 - x_2) = 0\\ \frac{\partial P}{\partial x_2} &= 4x_2 - 2\rho(1 - x_1 - x_2) = 0.\end{aligned}$$

These conditions give $x_1 = 2x_2$, and solving further yields

$x_2(\rho) = \dfrac{2\rho}{4 + 6\rho} = \dfrac{2\rho}{\rho\left(\frac{4}{\rho} + 6\right)}$ and thus $\lim_{\rho\to\infty} x_2(\rho) = x_2^* = \frac{1}{3}$, and also $x_1^* = \frac{2}{3}$.

Problem 5.3.1.4

Minimize $f(\mathbf{x}) = 2x_1^2 + x_2^2$ such that $g(\mathbf{x}) = 5 - x_1 + 3x_2 \leq 0$ by means of the *penalty function* approach.

Solution

The unconstrained solution, $x_1^* = x_2^* = 0$, violates the constraint, therefore it is active and P becomes $P = 2x_1^2 + x_2^2 + \rho(5 - x_1 + 3x_2)^2$ with the necessary conditions for an unconstrained minimum:

$$\begin{aligned}\frac{\partial P}{\partial x_1} &= 4x_1 - 2\rho(5 - x_1 + 3x_2) = 0\\ \frac{\partial P}{\partial x_2} &= 2x_2 + 6\rho(5 - x_1 + 3x_2) = 0.\end{aligned}$$

It follows that $x_2 = -6x_1$ and substituting into the first condition yields

$x_1(\rho) = \dfrac{10\rho}{4 + 38\rho} = \dfrac{10\rho}{\rho\left(\frac{4}{\rho} + 38\right)}$ and $\lim_{\rho\to\infty} x_1(\rho) = x_1^* = \frac{10}{38} = 0.2632$. This gives $x_2^* = -1.5789$ with $f(\mathbf{x}^*) = 2.6316$.

5.3.2 The Lagrangian method applied to equality constrained problems

Problem 5.3.2.1

Determine the minima and maxima of $f(\mathbf{x}) = x_1x_2$, such that $x_1^2+x_2^2 = 1$, by means of the Lagrangian method.

Solution

Here $L = x_1x_2 + \lambda(x_1^2 + x_2^2 - 1)$ and therefore the stationary conditions are

$$\frac{\partial L}{\partial x_1} = x_2 + 2x_1\lambda = 0, \qquad x_2 = -2x_1\lambda$$
$$\frac{\partial L}{\partial x_2} = x_1 + 2x_2\lambda = 0, \qquad x_1 = -2x_2\lambda.$$

From the equality it follows that $1 = x_1^2 + x_2^2 = 4x_1^2\lambda^2 + 4x_2^2\lambda^2 = 4\lambda^2$ giving $\lambda = \pm\frac{1}{2}$.

Choosing $\lambda = \frac{1}{2}$ gives $x_2 = -x_1$, $2x_1^2 = 1$, and $x_1 = \pm\frac{1}{\sqrt{2}}$.

This results in the possibilities:

$x_1 = \frac{1}{\sqrt{2}}$, $x_2 = -\frac{1}{\sqrt{2}} \Rightarrow f^* = -\frac{1}{2}$

or

$x_1 = -\frac{1}{\sqrt{2}}$, $x_2 = \frac{1}{\sqrt{2}} \Rightarrow f^* = -\frac{1}{2}$.

Alternatively choosing $\lambda = -\frac{1}{2}$ gives $x_2 = x_1$, $2x_1^2 = 1$, and $x_1 = \pm\frac{1}{\sqrt{2}}$ and the possibilities:

$x_1 = \frac{1}{\sqrt{2}}$, $x_2 = \frac{1}{\sqrt{2}} \Rightarrow f^* = \frac{1}{2}$

or

$x_1 = -\frac{1}{\sqrt{2}}$, $x_2 = -\frac{1}{\sqrt{2}} \Rightarrow f^* = \frac{1}{2}$.

These possibilities are sketched in Figure 5.7.

Problem 5.3.2.2

Determine the dimensions, radius r and height h, of the solid cylinder of minimum total surface area which can be cast from a solid metallic sphere of radius r_0.

Solution

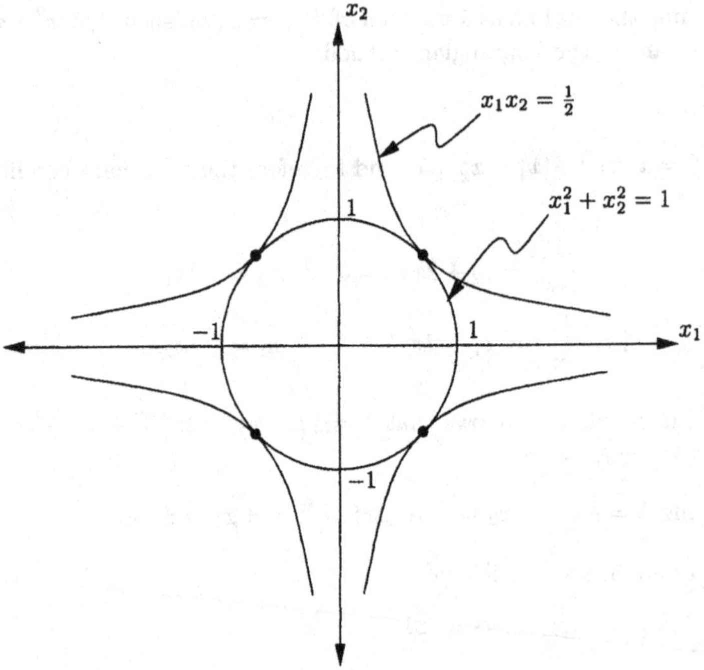

Figure 5.7

This problem is equivalent to: minimize $f(r,h) = 2\pi rh + 2\pi r^2$ such that $\pi r^2 h = \frac{4}{3}\pi r_0^3$. The Lagrangian is $L(r,h,\lambda) = -2\pi rh - 2\pi r^2 + \lambda(\pi r^2 h - \frac{4}{3}\pi r_0^3)$. The necessary conditions for stationary points are

$$\begin{aligned}
\frac{\partial L}{\partial r} &= -2\pi h - 4\pi r + \lambda 2\pi rh = 0 \\
\frac{\partial L}{\partial h} &= -2\pi r + \lambda \pi r^2 = 0 \\
\frac{\partial L}{\partial \lambda} &= \pi r^2 h - \tfrac{4}{3}\pi r_0^3 = 0.
\end{aligned}$$

The second condition gives $\lambda \pi r^2 = 2\pi r$, i.e. $\lambda = \frac{2}{r}$, and substituting in the first condition yields $r = \frac{h}{2}$. Substituting this value in the equality gives $\pi r^2 2r - \frac{4}{3}\pi r_0^3 = 0$, i.e.

$$r = r_0(\tfrac{2}{3})^{\frac{1}{3}}, \ h = 2r_0(\tfrac{2}{3})^{\frac{1}{3}} \text{ and } \lambda = \tfrac{2}{r}.$$

Problem 5.3.2.3

Minimize $f(\mathbf{x}) = -2x_1 - x_2 - 10$ such that $h(\mathbf{x}) = x_1 - 2x_2^2 - 3 = 0$. Show whether or not the candidate point, obtained via the Lagrangian method, is indeed a constrained *minimum*.

Solution

$$L(\mathbf{x}, \lambda) = -2x_1 - x_2 - 10 + \lambda(x_1 - 2x_2^2 - 3)$$

and the necessary stationary conditions are

$$\begin{aligned}\frac{\partial L}{\partial x_1} &= -2 + \lambda = 0,\\ \frac{\partial L}{\partial x_2} &= -1 - 4\lambda x_2 = 0\\ \frac{\partial L}{\partial \lambda} &= x_1 - 2x_2^2 - 3 = 0.\end{aligned}$$

Solving gives the candidate point $\lambda^* = 2$, $x_2^* = -\frac{1}{8}$, $x_1^* = 3.03$ with $f(\mathbf{x}^*) = -16.185$.

To prove that this point, $\mathbf{x}^* = [3.03, -\frac{1}{8}]^T$, indeed corresponds to a minimum requires the following argument. By Taylor, for any step $\mathbf{\Delta x}$ compatible with the constraint, i.e. $h(\mathbf{x}) = 0$, it follows that

$$f(\mathbf{x}^* + \mathbf{\Delta x}) = f(\mathbf{x}^*) + \nabla^T f(\mathbf{x}^*)\mathbf{\Delta x} \tag{5.3}$$

and $h(\mathbf{x}^* + \mathbf{\Delta x}) = h(\mathbf{x}^*) + \nabla^T h(\mathbf{x}^*)\mathbf{\Delta x} + \frac{1}{2}\mathbf{\Delta x}^T\nabla^2 h(\mathbf{x}^*)\mathbf{\Delta x} = 0$.

The latter equation gives $0 = [1, -4x_2]\begin{bmatrix}\Delta x_1\\ \Delta x_2\end{bmatrix} + \frac{1}{2}[\Delta x_1 \Delta x_2]\begin{bmatrix}0 & 0\\ 0 & -4\end{bmatrix}\begin{bmatrix}\Delta x_1\\ \Delta x_2\end{bmatrix}$, i.e.

$$\Delta x_1 = 4x_2\Delta x_2 + 2\Delta x_2^2 \tag{5.4}$$

and (5.3) gives

$$\Delta f = [-2 \ -1]\begin{bmatrix}\Delta x_1\\ \Delta x_2\end{bmatrix} = -2\Delta x_1 - \Delta x_2. \tag{5.5}$$

Substituting (5.4) into (5.5) results in $\Delta f = -2(4x_2\Delta x_2 + 2\Delta x_2^2) - \Delta x_2$ and setting $x_2 = x_2^* = \frac{1}{8}$ gives $\Delta f(\mathbf{x}^*) = \Delta x_2 - 4\Delta x_2^2 - \Delta x_2 = -4\Delta x_2^2 < 0$ for all Δx_2. Thus $\mathbf{x}^*$ is *not* a minimum, but in fact a constrained *maximum*.

Problem 5.3.2.4

Minimize $f(\mathbf{x}) = 3x_1^2 + x_2^2 + 2x_1x_2 + 6x_1 + 2x_2$ such that $h(\mathbf{x}) = 2x_1 - x_2 - 4 = 0$. Show that the candidate point you obtain is indeed a local constrained minimum.

Solution

The Lagrangian is given by

$$L(\mathbf{x}, \lambda) = 3x_1^2 + x_2^2 + 2x_1x_2 + 6x_1 + 2x_2 + \lambda(2x_1 - x_2 - 4)$$

and the associated necessary stationary conditions are

$$\begin{aligned} \frac{\partial L}{\partial x_1} &= 6x_1 + 2x_2 + 6 + 2\lambda = 0 \\ \frac{\partial L}{\partial x_2} &= 2x_2 + 2x_1 + 2 - \lambda = 0 \\ \frac{\partial L}{\partial \lambda} &= 2x_1 - x_2 - 4 = 0. \end{aligned}$$

Solving gives $x_1^* = \frac{7}{11}$, $x_2^* = -\frac{30}{11}$, $\lambda^* = -\frac{24}{11}$.

To prove that the above point indeed corresponds to a local minimum, the following further argument is required. Here

$$\boldsymbol{\nabla} f = \begin{bmatrix} 6x_1 + 2x_2 + 6 \\ 2x_2 + 2x_1 + 2 \end{bmatrix}, \ \boldsymbol{\nabla} h = \begin{bmatrix} 2 \\ -1 \end{bmatrix}, \ \boldsymbol{\nabla}^2 h = \mathbf{0} \text{ and } \mathbf{H} = \begin{bmatrix} 6 & 2 \\ 2 & 2 \end{bmatrix}$$

is positive-definite. Now for any step $\Delta\mathbf{x}$ compatible with the constraint $h(\mathbf{x}) = 0$ it follows that the changes in the constraint function and objective function are respectively given by

$$\Delta h = \boldsymbol{\nabla}^T h \Delta\mathbf{x} = [2 \ -1] \begin{bmatrix} \Delta x_1 \\ \Delta x_2 \end{bmatrix} = 2\Delta x_1 - \Delta x_2 = 0, \text{ i.e. } 2\Delta x_1 = \Delta x_2 \quad (5.6)$$

and

$$\Delta f = \boldsymbol{\nabla}^T f(\mathbf{x}^*)\Delta\mathbf{x} + \tfrac{1}{2}\Delta\mathbf{x}^T\mathbf{H}(\mathbf{x}^*)\Delta\mathbf{x}. \quad (5.7)$$

Also since $\boldsymbol{\nabla} f(\mathbf{x}^*) = \frac{24}{11}\begin{bmatrix} 2 \\ -1 \end{bmatrix}$, the first term in (5.7) may be written as

$$\boldsymbol{\nabla}^T f(\mathbf{x}^*)\Delta\mathbf{x} = \tfrac{24}{11}[2, -1]\begin{bmatrix} \Delta x_1 \\ \Delta x_2 \end{bmatrix} = \tfrac{24}{11}(2\Delta x_1 - \Delta x_2) = 0 \text{ (from (5.6))}.$$

Finally, substituting the latter expression into (5.7) gives, for any step $\Delta\mathbf{x}$ at $\mathbf{x}^*$ compatible with the constraint, the change in function value as $\Delta f = 0 + \frac{1}{2}\Delta\mathbf{x}^T\mathbf{H}(\mathbf{x}^*)\Delta\mathbf{x} > 0$ since $\mathbf{H}$ is positive-definite. The point $\mathbf{x}^*$ is therefore a strong local constrained minimum.

Problem 5.3.2.5

Maximize $f = xyz$ such that $\left(\frac{x}{a}\right)^2 + \left(\frac{y}{b}\right)^2 + \left(\frac{z}{c}\right)^2 = 1$.

Solution

$$L = xyz + \lambda\left(\left(\tfrac{x}{a}\right)^2 + \left(\tfrac{y}{b}\right)^2 + \left(\tfrac{z}{c}\right)^2 - 1\right)$$

with necessary conditions:

$$\begin{aligned}\frac{\partial L}{\partial x} &= yz + 2\lambda\frac{x}{a^2} = 0\\ \frac{\partial L}{\partial y} &= xz + 2\lambda\frac{y}{b^2} = 0\\ \frac{\partial L}{\partial z} &= xy + 2\lambda\frac{z}{c^2} = 0.\end{aligned}$$

Solving the above together with the given equality, yields $\lambda = -\frac{3}{2}xyz$, $x = \frac{1}{\sqrt{3}}a$, $y = \frac{1}{\sqrt{3}}b$ and $z = \frac{1}{\sqrt{3}}c$, and thus $f^* = \frac{1}{3\sqrt{3}}abc$.

Problem 5.3.2.6

Minimize $(x_1 - 1)^3 + (x_2 - 1)^2 + 2x_1x_2$ such that $x_1 + x_2 = 2$.

Solution

With $h(\mathbf{x}) = x_1 + x_2 - 2 = 0$ the Lagrangian is given by

$$L(\mathbf{x}, \lambda) = (x_1 - 1)^3 + (x_2 - 1)^2 + 2x_1x_2 + \lambda(x_1 + x_2 - 2)$$

with necessary conditions:

$$\begin{aligned}\frac{\partial L}{\partial x_1} = 3(x_1 - 1)^2 + 2x_2 + \lambda &= 0\\ \frac{\partial L}{\partial x_2} = 2(x_2 - 1) + 2x_1 + \lambda &= 0\\ x_1 + x_2 &= 2.\end{aligned}$$

Solving gives $\lambda^* = -2$ and the possible solutions

$$x_1^* = 1,\ x_2^* = 1 \text{ or } x_1^* = \tfrac{5}{3},\ x_2^* = \tfrac{1}{3}.$$

Analysis of the solutions:

For any $\mathbf{\Delta x}$ consistent with the constraint:

$$\Delta h = 0 = \boldsymbol{\nabla}^T h\mathbf{\Delta x} = [1,1]\begin{bmatrix}\Delta x_1\\ \Delta x_2\end{bmatrix},\ \text{i.e. } \Delta x_1 + \Delta x_2 = 0$$

and

$$\Delta f = \boldsymbol{\nabla}^T f\mathbf{\Delta x} + \tfrac{1}{2}\mathbf{\Delta x}\mathbf{H}(\overline{\mathbf{x}})\mathbf{\Delta x} \text{ where } \overline{\mathbf{x}} = \mathbf{x}^* + \theta\mathbf{\Delta x},\ 0 \leq \theta \leq 1.$$

For both candidate points $\mathbf{x}^*$ above, $\boldsymbol{\nabla}^T f = [2,2]$ and thus: $\boldsymbol{\nabla}^T f\mathbf{\Delta x} = 2(\Delta x_1 + \Delta x_2) = 0$. Considering only $\mathbf{\Delta x}\mathbf{H}(\overline{\mathbf{x}})\mathbf{\Delta x}$, it is clear that as $\mathbf{\Delta x} \to \mathbf{0}$, $\overline{\mathbf{x}} \to \mathbf{x}^*$, and $\Delta f > 0$ if $\mathbf{H}(\mathbf{x}^*)$ is positive-definite.

$\mathbf{H} = \begin{bmatrix} 6(x_1-1) & 2\\ 2 & 2\end{bmatrix}$ and thus at $\mathbf{x}^* = [1,1]^T$, $\mathbf{H} = \begin{bmatrix} 0 & 2\\ 2 & 2\end{bmatrix}$ is not positive-definite, and at $\mathbf{x}^* = \left[\frac{5}{3}, \frac{1}{3}\right]^T$, $\mathbf{H} = \begin{bmatrix} 4 & 2\\ 2 & 2\end{bmatrix}$ is positive-definite.

Therefore the point $\mathbf{x}^* = \left[\frac{5}{3}, \frac{1}{3}\right]^T$ is a strong local constrained minimum.

Problem 5.3.2.7

Determine the dimensions of a cylindrical can of maximum volume subject to the condition that the total surface area be equal to 24π. Show that the answer indeed corresponds to a maximum. (x_1 = radius, x_2 = height)

Solution

The problem is equivalent to:

minimize $f(x_1, x_2) = -\pi x_1^2 x_2$ such that $2\pi x_1^2 + 2\pi x_1 x_2 = A_0 = 24\pi$.

Thus with $h(\mathbf{x}) = 2\pi x_1^2 + 2\pi x_1 x_2 - 24\pi = 0$ the appropriate Lagrangian is

$$L(\mathbf{x}, \lambda) = -\pi x_1^2 x_2 + \lambda(2\pi x_1^2 + 2\pi x_1 x_2 - 24\pi)$$

with necessary conditions for a local minimum:

$$\begin{aligned}
\frac{\partial L}{\partial x_1} &= -2\pi x_1 x_2 + 4\pi\lambda x_1 + 2\pi\lambda x_2 = 0\\
\frac{\partial L}{\partial x_2} &= -\pi x_1^2 + 2\pi\lambda x_1 = 0\\
\text{and } h(\mathbf{x}) &= 2\pi x_1^2 + 2\pi x_1 x_2 - 24\pi = 0.
\end{aligned}$$

Solving gives $x_1^* = 2$, $x_2^* = 4$ with $\lambda^* = 1$ and $f^* = -16\pi$.

Analysis of the solution:

At $\mathbf{x}^*$ for change $\mathbf{\Delta x}$ compatible with the constraint it is required to show that

$$\Delta f = \nabla^T f \mathbf{\Delta x} + \tfrac{1}{2}\mathbf{\Delta x}^T \mathbf{H} \mathbf{\Delta x} > 0 \tag{5.8}$$

in the limit as $\mathbf{\Delta x} \to \mathbf{0}$ and $h(\mathbf{x}^* + \mathbf{\Delta x}) = 0$.

For $f(\mathbf{x})$ at $\mathbf{x}^*$ the gradient vector and Hessian are given by

$$\nabla f = \begin{bmatrix} -2\pi x_1 x_2 \\ -\pi x_1^2 \end{bmatrix} = -4\pi \begin{bmatrix} 4 \\ 1 \end{bmatrix} \text{ and}$$

$$\mathbf{H} = \begin{bmatrix} -2\pi x_2 & -2\pi x_1 \\ -2\pi x_1 & 0 \end{bmatrix} = \begin{bmatrix} -8\pi & -4\pi \\ -4\pi & 0 \end{bmatrix}.$$

For satisfaction of the constraint: $\Delta h = \nabla^T h \mathbf{\Delta x} + \tfrac{1}{2}\mathbf{\Delta x}^T \nabla^2 h \mathbf{\Delta x} = 0$, i.e.

$$\nabla^T h \mathbf{\Delta x} = -\tfrac{1}{2}\mathbf{\Delta x}^T \nabla^2 h \mathbf{\Delta x} \tag{5.9}$$

with $\nabla h = \begin{bmatrix} 4\pi x_1 + 2\pi x_2 \\ 2\pi x_1 \end{bmatrix} = 4\pi \begin{bmatrix} 4 \\ 1 \end{bmatrix}$

and where $\nabla^2 h = \begin{bmatrix} 4\pi & 2\pi \\ 2\pi & 0 \end{bmatrix}$. Now clearly $\nabla h = -\nabla f$ at the candidate point.

It therefore follows that $\nabla f^T \mathbf{\Delta x} = -\nabla^T h \mathbf{\Delta x} = \tfrac{1}{2}\mathbf{\Delta x}^T \nabla^2 h \mathbf{\Delta x}$.

Substituting in (5.8):

$$\begin{aligned}
\Delta f &= \tfrac{1}{2}\mathbf{\Delta x}^T(\nabla^2 h + \mathbf{H})\mathbf{\Delta x} \\
&= \tfrac{1}{2}\mathbf{\Delta x}^T \begin{bmatrix} -4\pi & -2\pi \\ -2\pi & 0 \end{bmatrix} \mathbf{\Delta x} = -\pi[\Delta x_1\ \Delta x_2] \begin{bmatrix} 2 & 1 \\ 1 & 0 \end{bmatrix} \begin{bmatrix} \Delta x_1 \\ \Delta x_2 \end{bmatrix} \\
&= -\pi[2\Delta x_1^2 + 2\Delta x_1 \Delta x_2].
\end{aligned}$$

From (5.9) in the limit as $\mathbf{\Delta x} \to \mathbf{0}$, $\Delta x_2 = -4\Delta x_1$, and thus

$\Delta f = -\pi[2\Delta x_1^2 + 2\Delta x_1(-4\Delta x_1)] = 6\pi \Delta x_1^2 > 0$, as expected for a constrained local minimum.

Problem 5.3.2.8

Minimize $f(\mathbf{x}) = x_1^2+x_2^2+\cdots+x_n^2$ such that $h(\mathbf{x}) = x_1+x_2+\cdots+x_n-1 = 0$.

Solution

The Lagrangian is

$$L = (x_1^2 + x_2^2 + \cdots + x_n^2) + \lambda(x_1 + x_2 + \cdots + x_n - 1)$$

with necessary conditions:

$$\frac{\partial L}{\partial x_i} = 2x_i + \lambda = 0, \; i = 1, \ldots, n.$$

Thus $\sum_{i=1}^{n} 2x_i + n\lambda = 0 \Rightarrow \lambda = -\frac{2}{n}$ and

$$2x_i - \tfrac{2}{n} = 0 \Rightarrow x_i = \tfrac{1}{n}, \; i = 1, 2, \ldots, n.$$

Therefore the distance $= \sqrt{f(\mathbf{x}^*)} = \sqrt{\frac{n}{n^2}} = \frac{1}{\sqrt{n}}$.

Test for a minimum at $\mathbf{x}^*$:

$$\Delta f = \boldsymbol{\nabla}^T f(\mathbf{x}^*)\boldsymbol{\Delta}\mathbf{x} + \tfrac{1}{2}\boldsymbol{\Delta}\mathbf{x}^T\mathbf{H}(\bar{\mathbf{x}})\boldsymbol{\Delta}\mathbf{x} \text{ and } \boldsymbol{\nabla}^T f = [2x_1, \ldots, 2x_n]^T$$

and $\mathbf{H}(\bar{\mathbf{x}}) = 2\begin{bmatrix} 1 & & & \\ & 1 & & \\ & & 1 & \\ & & & 1 \end{bmatrix}$ which is positive-definite, $\Delta f > 0$ if $\boldsymbol{\nabla} f(\mathbf{x}^*)\boldsymbol{\Delta}\mathbf{x} \geq 0$.

For $\boldsymbol{\Delta}\mathbf{x}$ such that $\Delta h = \boldsymbol{\nabla}^T h\boldsymbol{\Delta}\mathbf{x} = 0$, gives $\Delta x_1+\Delta x_2+\cdots+\Delta x_n = 0$.

Thus $\boldsymbol{\nabla}^T f(\mathbf{x}^*)\boldsymbol{\Delta}\mathbf{x} = 2[x_1^*, \ldots, x_n^*]^T\boldsymbol{\Delta}\mathbf{x} = \frac{2}{n}(\Delta x_1+\Delta x_2+\cdots+\Delta x_n) = 0$ and therefore $\mathbf{x}^*$ is indeed a constrained minimum.

Problem 5.3.2.9

Minimize $x_1^2 + x_2^2 + x_3^2$ such that $x_1 + 3x_2 + 2x_3 - 12 = 0$.

Solution

$$L = x_1^2 + x_2^2 + x_3^2 + \lambda(x_1 + 3x_2 + 2x_3 - 12)$$

$$\begin{aligned}
\frac{\partial L}{\partial x_1} = 2x_1 + \lambda &= 0\\
\frac{\partial L}{\partial x_2} = 2x_2 + 3\lambda &= 0\\
\frac{\partial L}{\partial x_3} = 2x_3 + 2\lambda &= 0\\
x_1 + 3x_2 + 2x_3 &= 12
\end{aligned}$$

with solution: $\lambda^* = -\frac{12}{7}$, $x_1^* = \frac{6}{7}$, $x_2^* = \frac{18}{7}$, $x_3^* = \frac{12}{7}$ with $f^* = 10.286$.

Test Δf for all $\mathbf{\Delta x}$ compatible with the constraint: $\Delta f = \nabla f^T \mathbf{\Delta x} + \frac{1}{2}\mathbf{\Delta x}^T\mathbf{H}\mathbf{\Delta x}$, $\nabla f = [2x_1, 2x_2, 2x_3]^T$, $\mathbf{H} = \begin{bmatrix} 2 & & 0 \\ & 2 & \\ 0 & & 2 \end{bmatrix}$ positive-definite and

$$\Delta h = \nabla^T h \mathbf{\Delta x} + \tfrac{1}{2}\mathbf{\Delta x}^T \nabla^2 h \mathbf{\Delta x} = 0, \Rightarrow \Delta x_1 + 3\Delta x_2 + 2\Delta x_3 = 0.$$

Therefore

$$\begin{aligned}
\Delta f(\mathbf{x}^*) &= 2x_1^*\Delta x_1 + 2x_2^*\Delta x_2 + 2x_3^*\Delta x_3 + \tfrac{1}{2}\mathbf{\Delta x}^T\mathbf{H}\mathbf{\Delta x}\\
&= 2\tfrac{6}{7}(\Delta x_1 + 3\Delta x_2 + 2\Delta x_3) + \tfrac{1}{2}\mathbf{\Delta x}^T\mathbf{H}\mathbf{\Delta x} = 0 + \tfrac{1}{2}\mathbf{\Delta x}^T\mathbf{H}\mathbf{\Delta x} > 0.
\end{aligned}$$

Thus $\mathbf{x}^*$ is indeed a local minimum.

5.3.3 Solution of inequality constrained problems via auxiliary variables

Problem 5.3.3.1

Consider the problem:

Minimize $f(\mathbf{x}) = x_1^2 + x_2^2 - 3x_1x_2$ such that $x_1^2 + x_2^2 - 6 \leq 0$.

Solution

Introducing an auxiliary variable θ the problem is transformed to an equality constrained problem with $x_1^2 + x_2^2 - 6 + \theta^2 = 0$. Since $\theta^2 \geq 0$ the original equality constraint is satisfied for all values of θ for which the new equality constraint is satisfied. Solve the new equality constrained problem with additional variable θ by the Lagrange method:

$$L(\mathbf{x}, \theta, \lambda) = x_1^2 + x_2^2 - 3x_1x_2 + \lambda(x_1^2 + x_2^2 - 6 + \theta^2)$$

with necessary conditions:

$$\begin{aligned}\frac{\partial L}{\partial x_1} &= 2x_1 - 3x_2 + 2\lambda x_1 = 0\\ \frac{\partial L}{\partial x_2} &= 2x_2 - 3x_1 + 2\lambda x_2 = 0\\ \frac{\partial L}{\partial \theta} &= 2\theta\lambda = 0\\ \frac{\partial L}{\partial \lambda} &= x_1^2 + x_2^2 - 6 + \theta^2 = 0.\end{aligned}$$

The third equation implies three possibilities:

(i) $\lambda = 0,\ \theta \neq 0,\ \Rightarrow x_1^* = x_2^* = 0,\ \theta^2 = 6$ and $f(\mathbf{x}) = 0$.

(ii) $\theta = 0,\ \lambda \neq 0$, then from the first two conditions:

$$2\lambda = -2 + \tfrac{3x_2}{x_1} = -2 + \tfrac{3x_1}{x_2} \Rightarrow x_1^2 = x_2^2.$$

Thus $x_1 = \pm x_2$ and from the last condition it follows that $x_1 = \pm\sqrt{3}$.

Choosing $x_1^* = -x_2^*$ gives $f(\mathbf{x}^*) = 15$, and choosing $x_1^* = x_2^* = \pm\sqrt{3}$ gives $f(\mathbf{x}^*) = -3$.

(iii) $\theta = 0$ and $\lambda = 0$ leads to a contradiction.

The constrained minimum therefore corresponds to case (ii) above.

Problem 5.3.3.2

Minimize $f(\mathbf{x}) = 2x_1^2 - 3x_2^2 - 2x_1$ such that $x_1^2 + x_2^2 \leq 1$.

Solution

Introduce auxiliary variable θ such that $x_1^2+x_2^2+\theta^2 = 1$. The Lagrangian is then

$$L(\mathbf{x}, \theta) = 2x_1^2 - 3x_2^2 - 2x_1 + \lambda(x_1^2 + x_2^2 + \theta^2 - 1)$$

and the necessary conditions for a minimum:

$$\frac{\partial L}{\partial x_1} = 4x_1 - 2 + 2\lambda x_1 = 0,\ \frac{\partial L}{\partial x_2} = -6x_2 + 2\lambda x_2 = 0,\ \frac{\partial L}{\partial \theta} = 2\lambda\theta = 0$$

$$\text{and } x_1^2 + x_2^2 + \theta^2 = 1.$$

The possibilities are:

either (i) $\lambda = 0$, $\theta \neq 0$ or (ii) $\lambda \neq 0$, $\theta = 0$ or (iii) $\lambda = 0$ and $\theta = 0$. Considering each possibility in turn gives the following:

(i) $\lambda = 0 \Rightarrow x_1^* = \frac{1}{2}$, $x_2^* = 0$, $\theta^{*2} = \frac{3}{4}$ and $f(x_1^*, x_2^*) = -\frac{1}{2}$.

(ii) $\theta = 0 \Rightarrow x_1^2 + x_2^2 = 1$; $4x_1 - 2 + 2\lambda x_1 = 0$; $x_2(-6 + 2\lambda) = 0$ giving firstly for $x_2^* = 0 \Rightarrow x_1^* = \pm 1$; $\lambda^* = -1$; -3 and $f^*(1, 0) = 0$; $f^*(-1, 0) = 4$, or secondly for $\lambda^* = 3$ it follows that $x_1^* = \frac{1}{5}$ and $x_2^* = \pm\sqrt{\frac{24}{25}}$, for which in both cases $f^* = -3.2$.

(iii) leads to a contradiction.

Inspection of the alternatives above gives the global minima at $x_1^* = \frac{1}{5}$ and $x_2^* = \pm\sqrt{\frac{24}{25}}$ with $f^* = -3.2$ and maximum at $x_1^* = -1$, $x_2^* = 0$, with $f^* = 4$.

Problem 5.3.3.3

Maximise $f(\mathbf{x}) = -x_1^2 + x_1x_2 - 2x_2^2 + x_1 + x_2$ such that $2x_1 + x_2 \leq 1$, $x_1, x_2 \geq 0$ by using auxiliary variables.

Solution

Here solve the minimization problem: $\min_{\mathbf{x}} f = x_1^2 - x_1x_2 + 2x_2^2 - x_1 - x_2$ such that $1 - 2x_1 - x_2 \geq 0$.

Introducing an auxiliary variable it follows that

$$L(\mathbf{x}, \theta) = x_1^2 - x_1x_2 + 2x_2^2 - x_1 - x_2 + \lambda(1 - 2x_1 - x_2 - \theta^2)$$

with necessary conditions:

$$\begin{aligned}
\frac{\partial L}{\partial x_1} &= 2x_1 - x_2 - 1 - 2\lambda = 0 \\
\frac{\partial L}{\partial x_2} &= -x_1 + 4x_2 - 1 - \lambda = 0 \\
\frac{\partial L}{\partial \theta} &= -2\lambda\theta = 0 \\
\frac{\partial L}{\partial \lambda} &= 1 - 2x_1 - x_2 - \theta^2 = 0.
\end{aligned}$$

The possibilities are:

(i) $\lambda = 0$ and $\theta \neq 0 \Rightarrow 2x_1 - x_2 - 1 = 0$ and $-x_1 + 4x_2 - 1 = 0$, giving $x_1 = \frac{5}{7}$ and $x_2 = \frac{3}{7}$. Substituting in the equality $1 - 2x_1 - x_2 - \theta^2 = 0 \Rightarrow \theta^2 = -\frac{6}{7} < 0$, which is not possible for θ real.

(ii) $\theta = 0$ and $\lambda \neq 0$ gives $1 - 2x_1 - x_2 = 0$ and solving together with the first two necessary conditions gives $x_2^* = \frac{3}{11}$, $x_1^* = \frac{4}{11}$ and $\lambda^* = \frac{1}{11}$.

5.3.4 Solution of inequality constrained problems via the Karush-Kuhn-Tucker conditions

Problem 5.3.4.1

Minimize $f(\mathbf{x}) = 3x_1 + x_2$

subject to $g(\mathbf{x}) = x_1^2 + x_2^2 - 5 \leq 0$.

Solution

For $L(\mathbf{x}, \lambda) = 3x_1 + x_2 + \lambda(x_1^2 + x_2^2 - 5)$ the KKT conditions are:

$$\begin{aligned}
\frac{\partial L}{\partial x_1} = 3 + 2\lambda x_1 &= 0\\
\frac{\partial L}{\partial x_2} = 1 + 2\lambda x_2 &= 0\\
x_1^2 + x_2^2 - 5 &\leq 0\\
\lambda(x_1^2 + x_2^2 - 5) &= 0\\
\lambda &\geq 0.
\end{aligned}$$

From $\lambda(x_1^2 + x_2^2 - 5) = 0$ it follows that either $\lambda = 0$ or $x_1^2 + x_2^2 - 5 = 0$.

If $\lambda = 0$ then the first two conditions are not satisfied and therefore $\lambda \neq 0$ and we have the equality $x_1^2 + x_2^2 - 5 = 0$. It now follows that $x_1 = -\frac{3}{2\lambda}$ and $x_2 = -\frac{1}{2\lambda}$.

Substituting these values into the equality yields

$$\left(-\tfrac{3}{2\lambda}\right)^2 + \left(-\tfrac{1}{2\lambda}\right)^2 = 5,$$

which implies that $\lambda^* = +\sqrt{\frac{1}{2}} > 0$. The optimal solution is thus $x_1^* = -\frac{3}{\sqrt{2}}$, $x_2^* = -\frac{1}{\sqrt{2}}$ with $f(\mathbf{x}^*) = -\frac{10}{\sqrt{2}}$.

Problem 5.3.4.2

Minimize $x_1^2+x_2^2-14x_1-6x_2-7$ such that $x_1+x_2 \leq 2$ and $x_1+2x_2 \leq 3$.

Solution

$$L(\mathbf{x}, \boldsymbol{\lambda}) = x_1^2 + x_2^2 - 14x_1 - 6x_2 - 7 + \lambda_1(x_1 + x_2 - 2) + \lambda_2(x_1 + 2x_2 - 3)$$

with KKT conditions:

$$\begin{aligned}
\frac{\partial L}{\partial x_1} = 2x_1 - 14 + \lambda_1 + \lambda_2 &= 0 \\
\frac{\partial L}{\partial x_2} = 2x_2 - 6 + \lambda_1 + 2\lambda_2 &= 0 \\
\lambda_1(x_1 + x_2 - 2) &= 0 \\
\lambda_2(x_1 + 2x_2 - 3) &= 0 \\
\boldsymbol{\lambda} &\geq 0.
\end{aligned}$$

The possibilities are:

(i) The choice $\lambda_1 \neq 0$ and $\lambda_2 \neq 0$ gives

$$x_1+x_2-2 = 0,\ x_1+2x_2-3 = 0 \ \Rightarrow\ x_2 = 1,\ x_1 = 1,\ \text{with } f = -25.$$

Both constraints are satisfied but $\lambda_2 = -8 < 0$ with $\lambda_2 = 20$.

(ii) $\lambda_1 \neq 0$ and $\lambda_2 = 0$ gives $x_1 + x_2 - 2 = 0$ and

$$\begin{aligned}
2x_1 - 14 + \lambda_1 &= 0 \\
2x_2 - 6 + \lambda_1 &= 0
\end{aligned}$$

which yield

$$\lambda_1 = 8 \geq 0 \text{ and } x_1 = 3,\ x_2 = -1,\ \text{with } f = -33$$

and both constraints are satisfied.

(iii) $\lambda_1 = 0$ and $\lambda_2 \neq 0$ gives $x_1 + 2x_2 - 3 = 0$ and it follows that $\lambda_2 = 4 \geq 0$ and $x_1 = 5$, $x_2 = -1$, with $f = -45$. However, the first constraint is violated.

(iv) The final possibility $\lambda_1 = \lambda_2 = 0$ gives $x_1 = 7$ and $x_2 = 3$ with $f = -65$ but both the first and second constraints are violated.

The unique optimum solution is therefore given by possibility (ii).

Problem 5.3.4.3

Minimize $f(\mathbf{x}) = x_1^2 + x_2^2 - 4x_1 - 6x_2 + 13$ such that $x_1 + x_2 \geq 7$ and $x_1 - x_2 \leq 2$.

Solution

We note that in this case the KKT conditions are also sufficient since the problem is convex.

$L(\mathbf{x}, \boldsymbol{\lambda}) = (x_1^2 + x_2^2 - 4x_1 - 6x_2 + 13) + \lambda_1(-x_1 - x_2 + 7) + \lambda_2(x_1 - x_2 - 2)$

with KKT conditions:

$$\begin{aligned}
\frac{\partial L}{\partial x_1} = 2x_1 - 4 - \lambda_1 + \lambda_2 &= 0 \\
\frac{\partial L}{\partial x_2} = 2x_2 - 6 - \lambda_1 - \lambda_2 &= 0 \\
\lambda_1(-x_1 - x_2 + 7) &= 0 \\
\lambda_2(x_1 - x_2 - 2) &= 0 \\
\boldsymbol{\lambda} &\geq 0.
\end{aligned}$$

The possibilities are:

(i) $\lambda_1 = \lambda_2 = 0 \Rightarrow x_1 = 2,\ x_2 = 3$ with $x_1 + x_2 = 2 + 3 < 7$ and thus the first constraint is not satisfied.

(ii) $\lambda_1 \neq 0,\ \lambda_2 = 0 \Rightarrow 2x_1 - 4 - \lambda_1 = 0,\ 2x_2 - 6 - \lambda_1 = 0$ and $x_1 - x_2 + 1 = 0$.

The above implies $-x_1 - x_2 + 7 = 0,\ \lambda_2 = 0,\ \lambda_1 = 2,\ x_2 = 4,\ x_1 = 3$ with $f = 2$. This point satisfies all the KKT conditions and is therefore a local minimum.

(iii) $\lambda_1 = 0$ and $\lambda_2 \neq 0 \Rightarrow 2x_1 - 4 + \lambda_2 = 0,\ 2x_2 - 6 - \lambda_2 = 0,\ 2x_1 + 2x_2 - 10 = 0,\ x_1 - x_2 - 2 = 0$ and solving yields

$$x_1 = \tfrac{7}{2},\ \lambda_1 = 0,\ x_2 = \tfrac{3}{2} \text{ and } \lambda_2 = -3.$$

This point violates the condition that $\lambda_2 \geq 0$, and is therefore not a local minimum.

(iv) $\lambda_1 \neq 0$ and $\lambda_2 \neq 0 \Rightarrow 2x_1 - 4 - \lambda_1 + \lambda_2 = 0$, $2x_2 - 6 - \lambda_1 - \lambda_2 = 0$, and $-x_1 - x_2 + 7 = 0$, and $x_1 - x_2 - 2 = 0 \Rightarrow x_2 = \frac{5}{2}$, $x_1 = \frac{9}{2}$ with $\lambda_1 = 2$ and $\lambda_2 = -3 < 0$ which violates the condition $\lambda \geq 0$.

The unique optimum solution therefore corresponds to possibility (ii), i.e. $x_1^* = 3$, $\lambda_1^* = 2$, $x_2^* = 4$, $\lambda_2^* = 0$ with $f^* = 2$.

Problem 5.3.4.4

Minimize $x_1^2 + 2(x_2 + 1)^2$ such that $-x_1 + x_2 = 2$, $-x_1 - x_2 - 1 \leq 0$.

Solution

Here the Lagrangian is given by

$$L(\mathbf{x}, \lambda, \mu) = x_1^2 + 2(x_2 + 1)^2 + \lambda(-x_1 - x_2 - 1) + \mu(-x_1 + x_2 - 2)$$

with KKT conditions:

$$\begin{aligned}
\frac{\partial L}{\partial x_1} = 2x_1 - \lambda - \mu &= 0 \\
\frac{\partial L}{\partial x_2} = 4(x_2 + 1) - \lambda + \mu &= 0 \\
-x_1 + x_2 - 2 &= 0 \\
-x_1 - x_2 - 1 &\leq 0 \\
\lambda(-x_1 - x_2 - 1) &= 0 \\
\lambda &\geq 0.
\end{aligned}$$

Possibilities:

(i) $\lambda \neq 0 \Rightarrow -x_1 - x_2 - 1 = 0$ and with $-x_1 + x_2 - 2 = 0 \Rightarrow x_1 = -\frac{3}{2}$, $x_2 = \frac{1}{2}$. Substituting into the first two conditions give $\lambda = \frac{3}{2}$ and $\mu = -\frac{9}{2}$, and thus all conditions are satisfied.

(ii) $\lambda = 0 \Rightarrow 2x_1 - \mu = 0$ and $4x_2 + 4 + \mu = 0$ giving $2x_1 + 4x_2 + 4 = 0$ and with $-x_1 + x_2 - 2 = 0$ it follows that $x_2 = 0$ and $x_1 = -2$. However, $-x_1 - x_2 - 1 = 1$ which does not satisfy the inequality.

Case (i) therefore represents the optimum solution with $f^* = \frac{27}{4}$.

Problem 5.3.4.5

Determine the shortest distance from the origin to the set defined by:

$$4 - x_1 - x_2 \le 0, \quad 5 - 2x_1 - x_2 \le 0.$$

Solution

$$L(\mathbf{x}, \boldsymbol{\lambda}) = x_1^2 + x_2^2 + \lambda_1(4 - x_1 - x_2) + \lambda_2(5 - 2x_1 - x_2)$$

with KKT conditions:

$$\begin{aligned} \frac{\partial L}{\partial x_1} = 2x_1 - \lambda_1 - 2\lambda_2 &= 0 \\ \frac{\partial L}{\partial x_2} = 2x_2 - \lambda_1 - \lambda_2 &= 0 \\ \lambda_1 g_1 + \lambda_2 g_2 &= 0 \\ \lambda_1, \lambda_2 &\ge 0. \end{aligned}$$

The possibilities are:

(i) $\lambda_1 = \lambda_2 = 0 \;\Rightarrow\; x_1 = x_2 = 0$ and both constraints are violated.

(ii) $\lambda_1 = 0, \lambda_2 \neq 0 \;\Rightarrow\; 5 - 2x_1 - x_2 = 0$ which gives $x_2 = 1$, $x_1 = 2$ and $\lambda_2 = 2 > 0$, but this violates constraint g_1.

(iii) $\lambda_1 \neq 0, \lambda_2 = 0 \;\Rightarrow\; x_1 = x_2 = 2$ and $\lambda_1 = 4 > 0$.

(iv) $\lambda_1 \neq 0, \lambda_2 \neq 0 \;\Rightarrow\; g_1 = 0$ and $g_2 = 0$ which implies that $\lambda_2 = -4 < 0$, $x_1 = 1$, $x_2 = 3$ which violates the non-negativity condition on λ_2.

The solution $\mathbf{x}^*$ therefore corresponds to case (iii) with shortest distance $= \sqrt{8}$.

Problem 5.3.4.6

Minimize $x_1^2 + x_2^2 - 2x_1 - 2x_2 + 2$

such that $-2x_1 - x_2 + 4 \le 0$ and $-x_1 - 2x_2 + 4 \le 0$.

Solution

$L(\mathbf{x}, \boldsymbol{\lambda}) = x_1^2 + x_2^2 - 2x_1 - 2x_2 + 2 + \lambda_1(-2x_1 - x_2 + 4) + \lambda_2(-x_1 - 2x_2 + 4)$

with KKT conditions:

$$\begin{aligned}
\frac{\partial L}{\partial x_1} = 2x_1 - 2 - 2\lambda_1 - \lambda_2 &= 0 \\
\frac{\partial L}{\partial x_2} = 2x_2 - 2 - \lambda_1 - 2\lambda_2 &= 0 \\
g_1 = -2x_1 - x_2 + 4 &\le 0 \\
g_2 = -x_1 - 2x_2 + 4 &\le 0 \\
\lambda_1 g_1 = 0; \ \lambda_2 g_2 &= 0 \\
\lambda_1 \ge 0; \ \lambda_2 &\ge 0.
\end{aligned}$$

The fifth conditions give the following possibilities:

(i) $\lambda_1 = \lambda_2 = 0 \Rightarrow 2x_1 - 2 = 0, 2x_2 - 2 = 0, x_1 = x_2 = 1$, not valid since both g_1 and $g_2 > 0$.

(ii) $\lambda_1 = 0$ and $g_2 = 0 \Rightarrow 2x_1 - 2 - \lambda_2 = 0, 2x_2 - 2 - 2\lambda_2 = 0$ and $-x_1 - 2x_2 + 4 = 0$ which yield

$$x_1 = \tfrac{6}{5}, \ x_2 = \tfrac{7}{5}, \ \lambda_2 = \tfrac{2}{5} > 0$$

but, not valid since $g_1 = 0.2 > 0$.

(iii) $g_1 = 0, \lambda_2 = 0 \Rightarrow 2x_1 - 2 - 2\lambda_1 = 0, 2x_2 - 2 - \lambda_1 = 0$ which together with $g_1 = 0$ give $x_1 = \frac{7}{5}, x_2 = \frac{6}{5}$ and $\lambda_1 = \frac{2}{5} > 0$, but not valid since $g_2 = 0.2 > 0$.

(iv) $g_1 = g_2 = 0 \Rightarrow x_1 = x_2 = \frac{4}{3}, \lambda_1 = \lambda_2 = \frac{2}{9} > 0$.

Since (iv) satisfies all the conditions it corresponds to the optimum solution with $f(\mathbf{x}^*) = \frac{2}{9}$.

Problem 5.3.4.7

Minimize $\left(x_1 - \frac{9}{4}\right)^2 + (x_2 - 2)^2$

such that $g_1(\mathbf{x}) = x_1^2 - x_2 \le 0$, $g_2(\mathbf{x}) = x_1 + x_2 - 6 \le 0$, and $x_1, x_2 \ge 0$.

Is the point $\overline{\mathbf{x}} = \left(\frac{3}{2}, \frac{9}{4}\right)$ a local minimum?

Solution

$L(\mathbf{x}, \boldsymbol{\lambda}) = \left(x_1 - \frac{9}{4}\right)^2 + (x_2-2)^2 + \lambda_1(x_1^2 - x_2) + \lambda_2(x_1+x_2-6) - \lambda_3 x_1 - \lambda_4 x_2$

with KKT conditions:

$$\begin{aligned}
\frac{\partial L}{\partial x_1} = 2\left(x_1 - \tfrac{9}{4}\right) + 2\lambda_1 x_1 + \lambda_2 - \lambda_3 &= 0\\
\frac{\partial L}{\partial x_2} = 2(x_2 - 2) - \lambda_1 + \lambda_2 - \lambda_4 &= 0\\
g_1 \le 0 \text{ and } g_2 &\le 0\\
-x_1 \le 0 \text{ and } -x_2 &\le 0\\
\lambda_1(x_1^2 - x_2) = 0;\ \ \lambda_2(x_1 + x_2 - 6) &= 0\\
\lambda_3 x_1 = 0;\ \ \lambda_4 x_2 &= 0\\
\lambda_1 \ge \lambda_2 \ge \lambda_3 \ge \lambda_4 &\ge 0.
\end{aligned}$$

At $\overline{\mathbf{x}} = \left(\frac{3}{2}, \frac{9}{4}\right)$, x_1 and $x_2 \neq 0 \ \Rightarrow\ \lambda_3 = \lambda_4 = 0$ and $g_1 = \left(\frac{3}{2}\right)^2 - \frac{9}{4} = 0$ and $g_2 = \frac{3}{2} + \frac{9}{4} - 6 = \frac{15}{4} - 6 < 0$ and thus $\lambda_2 = 0$.

Also from the first condition it follows that since $2\left(\frac{3}{2} - \frac{9}{4}\right) + 2\lambda_1 \frac{3}{2} = 0$ that $\lambda_1 = \frac{1}{2} > 0$. This value also satisfies the second condition.

Since $\overline{\mathbf{x}}$ satisfies all the KKT conditions, and all the constraints are convex, it is indeed the global constrained optimum.

Problem 5.3.4.8

Minimize $x_1^2 + x_2^2 - 8x_1 - 10x_2$ such that $3x_1 + 2x_2 - 6 \le 0$.

Solution

$$L(\mathbf{x}, \lambda) = x_1^2 + x_2^2 - 8x_1 - 10x_2 + \lambda(3x_1 + 2x_2 - 6)$$

with KKT conditions:

$$\begin{aligned}
\frac{\partial L}{\partial x_1} = 2x_1 - 8 + 3\lambda &= 0\\
\frac{\partial L}{\partial x_2} = 2x_2 - 10 + 2\lambda &= 0\\
\lambda(3x_1 + 2x_2 - 6) &= 0.
\end{aligned}$$

The possibilities are:

(i) $\lambda = 0$ then $x_1 = 4$ and $x_2 = 5$ giving $3x_1 + 2x_2 - 6 = 16 > 0$ and thus this point is not valid.

(ii) $3x_1 + 2x_2 - 6 = 0$ then $\lambda = \frac{32}{13}$ and $x_1 = \frac{4}{13}$, $x_2 = \frac{33}{13}$.

Thus the solution corresponds to case (ii).

Problem 5.3.4.9

Minimize $f(\mathbf{x}) = x_1^2 - x_2$ such that $x_1 \geq 1$, $x_1^2 + x_2^2 \leq 26$.

Solution

$$L(\mathbf{x}, \boldsymbol{\lambda}) = x_1^2 - x_2 + \lambda_1(1 - x_1) + \lambda_2(x_1^2 + x_2^2 - 26)$$

with KKT conditions:

$$\begin{aligned}
\frac{\partial L}{\partial x_1} = 2x_1 - \lambda_1 + 2\lambda_2 x_2 &= 0 \\
\frac{\partial L}{\partial x_2} = -1 + 2\lambda_2 x_2 &= 0 \\
\lambda_1(1 - x_1) &= 0 \\
\lambda_2(x_1^2 + x_2^2 - 26) &= 0 \\
x_1^2 + x_2^2 - 26 &\leq 0 \\
1 - x_1 &\leq 0 \\
\lambda_1 \geq 0; \quad \lambda_2 &\geq 0.
\end{aligned}$$

Investigate the possibilities implied by the third and fourth conditions.

By inspection the possibility $1 - x_1 = 0$ and $x_1^2 + x_2^2 - 26 = 0$ yields $x_1 = 1$; $x_2 = 5$. This gives $\lambda_2 = \frac{1}{10} > 0$ and $\lambda_1 = 2 > 0$ which satisfies the last condition. Thus all the conditions are now satisfied. Since $f(\mathbf{x})$ and all the constraints functions are convex the conditions are also sufficient and $\mathbf{x}^* = [1, 5]^T$ with $f(\mathbf{x}^*) = -4$ is a constrained minimum.

5.3.5 Solution of constrained problems via the dual problem formulation

Problem 5.3.5.1

Minimize $x_1^2 + 2x_2^2$ such that $x_2 \geq -x_1 + 2$.

Solution

The constraint in standard form is $2 - x_1 - x_2 \leq 0$ and the Lagrangian therefore:

$$L(\mathbf{x}, \lambda) = x_1^2 + 2x_2^2 + \lambda(2 - x_1 - x_2).$$

For a given value of λ the stationary conditions are

$$\begin{aligned} \frac{\partial L}{\partial x_1} = 2x_1 - \lambda &= 0 \\ \frac{\partial L}{\partial x_2} = 4x_2 - \lambda &= 0 \end{aligned}$$

giving $x_1 = \frac{\lambda}{2}$ and $x_2 = \frac{\lambda}{4}$. Since the Hessian of L, $\mathbf{H}_L$ is positive-definite the solution corresponds to a minimum.

Substituting the solution into L gives the dual function:

$$h(\lambda) = \tfrac{\lambda^2}{4} + \tfrac{\lambda^2}{8} + \lambda\left(2 - \tfrac{\lambda}{2} - \tfrac{\lambda}{4}\right) = -\tfrac{3}{8}\lambda^2 + 2\lambda.$$

Since $\frac{d^2h}{d\lambda^2} < 0$ the maximum occurs where $\frac{dh}{d\lambda} = -\frac{6}{8}\lambda + 2 = 0$, i.e. $\lambda^* = \frac{8}{3}$ with

$$h(\lambda^*) = -\tfrac{3}{8}\left(\tfrac{8}{3}\right)^2 + 2\left(\tfrac{8}{3}\right) = \tfrac{8}{3}.$$

Thus $\mathbf{x}^* = \left(\frac{4}{3}, \frac{2}{3}\right)^T$ with $f(\mathbf{x}^*) = \frac{8}{3}$.

Problem 5.3.5.2

Minimize $x_1^2 + x_2^2$ such that $2x_1 + x_2 \leq -4$.

Solution

$$L(\mathbf{x}, \lambda) = x_1^2 + x_2^2 + \lambda(2x_1 + x_2 + 4)$$

and the necessary conditions for a minimum with respect to $\mathbf{x}$ imply

$$\begin{aligned} \frac{\partial L}{\partial x_1} = 2x_1 + 2\lambda &= 0 \\ \text{and } \frac{\partial L}{\partial x_2} = 2x_2 + \lambda &= 0 \end{aligned}$$

giving $x_1 = -\lambda$ and $x_2 = -\frac{\lambda}{2}$. Since $\mathbf{H}_L$ is positive-definite the solution is indeed a minimum with respect to $\mathbf{x}$. Substituting in L gives

$$h(\lambda) = -\tfrac{5}{4}\lambda^2 + 4\lambda.$$

Since $\frac{d^2h}{d\lambda^2} = -\frac{5}{2} < 0$ the maximum occurs where $\frac{dh}{d\lambda} = -\frac{5}{2}\lambda + 4 = 0$, i.e. where $\lambda^* = \frac{8}{5} > 0$ and $f(\mathbf{x}^*) = h(\lambda^*) = \frac{16}{5}$.

The solution is thus $x_1^* = -\frac{8}{5}$; $x_2^* = -\frac{4}{5}$ which satisfies the KKT conditions.

Problem 5.3.5.3

Minimize $2x_1^2 + x_2^2$ such that $x_1 + 2x_2 \geq 1$.

Solution

$$L(\mathbf{x}, \lambda) = 2x_1^2 + x_2^2 + \lambda(1 - x_1 - 2x_2)$$

with stationary conditions

$$\begin{aligned} \frac{\partial L}{\partial x_1} = 4x_1 - \lambda &= 0 \\ \frac{\partial L}{\partial x_2} = 2x_2 - 2\lambda &= 0 \end{aligned}$$

giving $x_1 = \frac{\lambda}{4}$ and $x_2 = \lambda$. Since $\mathbf{H}_L$ is positive definite the solution is a minimum and

$$h(\lambda) = \tfrac{1}{8}\lambda^2 + \lambda^2 + \lambda - \tfrac{1}{4}\lambda^2 - 2\lambda^2 = -\tfrac{9}{8}\lambda^2 + \lambda.$$

Since $\frac{d^2h}{d\lambda^2} < 0$ the maximum occurs where $\frac{dh}{d\lambda} = -\frac{9}{4}\lambda + 1 = 0$, i.e. $\lambda = \frac{4}{9}$ and since $\lambda > 0$, $\lambda \in D$.

Thus

$$\begin{aligned} h(\lambda^*) &= -\tfrac{9}{8}\left(\tfrac{4}{9}\right)^2 + \tfrac{4}{9} = \tfrac{2}{9} \\ \text{and } x_1^* &= \tfrac{1}{4}\left(\tfrac{4}{9}\right) = \tfrac{1}{9},\ x_2^* = \tfrac{4}{9}. \end{aligned}$$

Test: $f(\mathbf{x}^*) = 2\left(\frac{1}{9}\right)^2 + \left(\frac{4}{9}\right)^2 = \frac{2}{81} + \frac{16}{81} = \frac{18}{81} = \frac{2}{9} = h(\lambda^*)$.

Problem 5.3.5.4

Minimize $(x_1 - 1)^2 + (x_2 - 2)^2$ such that $x_1 - x_2 \leq 1$ and $x_1 + x_2 \leq 2$.

Solution

$$L(\mathbf{x}, \boldsymbol{\lambda}) = (x_1 - 1)^2 + (x_2 - 2)^2 + \lambda_1(x_1 - x_2 - 1) + \lambda_2(x_1 + x_2 - 2)$$

and it follows that for a fixed choice of $\boldsymbol{\lambda} = [\lambda_1, \lambda_2]^T$ the stationary conditions are:

$$\begin{aligned}\frac{\partial L}{\partial x_1} &= 2(x_1 - 1) + \lambda_1 + \lambda_2 = 0\\ \frac{\partial L}{\partial x_2} &= 2(x_2 - 2) - \lambda_1 + \lambda_2 = 0\end{aligned}$$

which give $x_1 = 1 - \frac{1}{2}(\lambda_1 + \lambda_2)$ and $x_2 = 2 + \frac{1}{2}(\lambda_1 - \lambda_2)$. $\mathbf{H}_L = \begin{bmatrix} 2 & 0 \\ 0 & 2 \end{bmatrix}$ is positive-definite and therefore the solution is a minimum with respect to $\mathbf{x}$. Substituting the solution into L gives

$$h(\boldsymbol{\lambda}) = -\tfrac{\lambda_1^2}{2} - \tfrac{\lambda_2^2}{2} + \lambda_2.$$

The necessary conditions for a maximum are

$$\begin{aligned}\frac{\partial h}{\partial \lambda_1} &= -\tfrac{2\lambda_1}{2} = 0, \quad \text{i.e. } \lambda_1^* = 0\\ \frac{\partial h}{\partial \lambda_2} &= -\tfrac{2\lambda_2}{2} + 1 = 0, \quad \text{i.e. } \lambda_2^* = 1\end{aligned}$$

and since the Hessian of h with respect to $\boldsymbol{\lambda}$ is given by $\mathbf{H}_h = \begin{bmatrix} -1 & 0 \\ 0 & -1 \end{bmatrix}$ which is negative-definite the solution indeed corresponds to a maximum, with $h(\boldsymbol{\lambda}^*) = \frac{1}{2} = f(\mathbf{x}^*)$ and thus $x_1^* = \frac{1}{2}$ and $x_2^* = \frac{3}{2}$.

5.3.6 Quadratic programming problems

Problem 5.3.6.1

Minimize $f(\mathbf{x}) = x_1^2 + x_2^2 + x_3^2$ such that $h_1(\mathbf{x}) = x_1 + x_2 + x_3 = 0$ and $h_2(\mathbf{x}) = x_1 + 2x_2 + 3x_3 - 1 = 0$.

Solution

Here, for the equality constrained problem the solution is obtained via the Lagrangian method with

$$L(\mathbf{x}, \boldsymbol{\lambda}) = x_1^2 + x_2^2 + x_3^2 + \lambda_1(x_1 + x_2 + x_3) + \lambda_2(x_1 + 2x_2 + 3x_3 - 1).$$

The necessary conditions for a minimum give:

$$\frac{\partial L}{\partial x_1} = 2x_1 + \lambda_1 + \lambda_2 = 0, \qquad x_1 = -\tfrac{1}{2}(\lambda_1 + \lambda_2)$$
$$\frac{\partial L}{\partial x_2} = 2x_2 + \lambda_1 + 2\lambda_2 = 0, \qquad x_2 = -\tfrac{1}{2}(\lambda_1 + 2\lambda_2)$$
$$\frac{\partial L}{\partial x_3} = 2x_3 + \lambda_1 + 3\lambda_2 = 0, \qquad x_3 = -\tfrac{1}{2}(\lambda_1 + 3\lambda_2).$$

Substituting into the equality constraints gives:

$$-\left(\tfrac{\lambda_1+\lambda_2}{2} + \tfrac{\lambda_1+2\lambda_2}{2} + \tfrac{\lambda_1+3\lambda_2}{2}\right) = 0, \text{ i.e. } \lambda_1 + 2\lambda_2 = 0$$

and

$$\tfrac{1}{2}(\lambda_1 + \lambda_2) + (\lambda_1 + 2\lambda_2) + \tfrac{3}{2}(\lambda_1 + 3\lambda_2) = -1, \text{ i.e. } 3\lambda_1 + 7\lambda_2 = -1.$$

Solving for the λ's: $\lambda_2 = -1$ and $\lambda_1 = 2$.

The candidate solution is therefore: $x_1^* = -\frac{1}{2}$, $x_2^* = 0$, $x_3^* = \frac{1}{2}$.

For the further analysis:

$$f(\mathbf{x} + \mathbf{\Delta x}) = f(\mathbf{x}) + \boldsymbol{\nabla}^T f \mathbf{\Delta x} + \tfrac{1}{2}\mathbf{\Delta x H \Delta x}$$

where $\boldsymbol{\nabla} f = (2x_1, 2x_2, 2x_3)^T$ and thus $\boldsymbol{\nabla} f(\mathbf{x}^*) = [-1, 0, 1]^T$ and $\mathbf{H} = \begin{bmatrix} 2 & 0 & 0 \\ 0 & 2 & 0 \\ 0 & 0 & 2 \end{bmatrix}$ is positive-definite.

For changes consistent with the constraints:

$$0 = \Delta h_1 = \boldsymbol{\nabla}^T h_1 \mathbf{\Delta x} + \tfrac{1}{2}\mathbf{\Delta x}^T \boldsymbol{\nabla}^2 h_1 \mathbf{\Delta x}, \quad \text{with} \quad \boldsymbol{\nabla} h_1 = \begin{bmatrix} 1 \\ 1 \\ 1 \end{bmatrix}$$

$$0 = \Delta h_2 = \boldsymbol{\nabla}^T h_2 \mathbf{\Delta x} + \tfrac{1}{2}\mathbf{\Delta x}^T \boldsymbol{\nabla}^2 h_2 \mathbf{\Delta x}, \quad \text{with} \quad \boldsymbol{\nabla} h_2 = \begin{bmatrix} 1 \\ 2 \\ 3 \end{bmatrix}.$$

It follows that $\Delta x_1 + \Delta x_2 + \Delta x_3 = 0$ and $\Delta x_1 + 2\Delta x_2 + 3\Delta x_3 = 0$ giving $-\Delta x_1 + \Delta x_3 = 0$ and thus

$$\Delta f(\mathbf{x}^*) = -\Delta x_1 + \Delta x_3 + \tfrac{1}{2}\mathbf{\Delta x}^T \mathbf{H} \mathbf{\Delta x} = \tfrac{1}{2}\mathbf{\Delta x}^T \mathbf{H} \mathbf{\Delta x} \geq 0.$$

The candidate point $\mathbf{x}^*$ above is therefore a constrained minimum.

Problem 5.3.6.2

Minimize $f(\mathbf{x}) = -2x_1 - 6x_2 + x_1^2 - 2x_1x_2 + 2x_2^2$ such that $x_1 \geq 0$, $x_2 \geq 0$ and $g_1(\mathbf{x}) = x_1 + x_2 \leq 2$ and $g_2(\mathbf{x}) = -x_1 + 2x_2 \leq 2$.

Solution

In matrix form the problem is:

minimize $f(\mathbf{x}) = \frac{1}{2}\mathbf{x}^T\mathbf{A}\mathbf{x} + \mathbf{b}^T\mathbf{x}$ where $\mathbf{A} = \begin{bmatrix} 2 & -2 \\ -2 & 4 \end{bmatrix}$, $\mathbf{b} = \begin{bmatrix} -2 \\ -6 \end{bmatrix}$ subject to the specified linear constraints.

First determine the unconstrained minimum:

$$\mathbf{x}^* = -\mathbf{A}^{-1}\mathbf{b} = -\tfrac{1}{4}\begin{bmatrix} 4 & 2 \\ 2 & 2 \end{bmatrix}\begin{bmatrix} -2 \\ -6 \end{bmatrix} = \begin{bmatrix} 5 \\ 4 \end{bmatrix}.$$

Test for violation of constraints:

$$x_1, x_2 > 0, \quad x_1 + x_2 = 9 > 2; \quad -x_1 + 2x_2 = -5 + 8 = 3 > 2.$$

Thus two constraints are violated. Considering each separately active assume firstly that $x_1 + x_2 = 2$. For this case $L = f(\mathbf{x}) + \lambda g_1(\mathbf{x}) = \frac{1}{2}\mathbf{x}^T\mathbf{A}\mathbf{x} + \mathbf{b}^T\mathbf{x} + \lambda(x_1 + x_2 - 2)$. The necessary conditions are:

$$\begin{bmatrix} \mathbf{A} & & 1 \\ & & 1 \\ 1 & 1 & 0 \end{bmatrix}\begin{bmatrix} \mathbf{x} \\ \lambda \end{bmatrix} = \begin{bmatrix} -\mathbf{b} \\ 2 \end{bmatrix}, \text{ i.e. solve } \begin{bmatrix} 2 & -2 & 1 \\ -2 & 4 & 1 \\ 1 & 1 & 0 \end{bmatrix}\begin{bmatrix} x_1 \\ x_2 \\ \lambda \end{bmatrix} = \begin{bmatrix} 2 \\ 6 \\ 2 \end{bmatrix}.$$

This, by Cramer's rule, gives $x_1 = \frac{D_1}{D} = \frac{4}{5}$; $x_2 = \frac{D_2}{D} = \frac{6}{5}$ and $\lambda = \frac{D_3}{D} = \frac{14}{5} > 0$, and therefore

$$x_1,\ x_2 > 0 \text{ and } -x_1 + 2x_2 = -\tfrac{4}{5} + \tfrac{2.6}{5} = \tfrac{8}{5} < 2.$$

Thus with all the constraints satisfied and $\lambda > 0$ the KKT sufficient conditions apply and $\mathbf{x}^* = [\frac{4}{5}; \frac{6}{5}]$ is a local constrained minimum. Indeed, since the problem is convex it is in fact the global minimum and no further investigation is required.

Problem 5.3.6.3

Minimize $f(\mathbf{x}) = x_1^2 + 4x_2^2 - 2x_1 + 8x_2$ such that $5x_1 + 2x_2 \leq 4$ and $x_1, x_2 \geq 0$.

Solution

Try various possibilities and test for satisfaction of the KKT conditions.

(i) For the unconstrained solution: solve $\boldsymbol{\nabla} f(\mathbf{x}) = \mathbf{0}$, i.e. $2x_1 - 2 = 0$, $8x_2 + 8 = 0$ giving $x_1 = 1$; $x_2 = -1$ which violates the second non-negativity constraint.

(ii) Setting $x_2 = 0$ results in $x_1 = 1$ which violates the constraint $5x_1 + 2x_2 \leq 4$.

(iii) Similarly, setting $x_1 = 0$ results in $x_2 = -1$, which violates the second non-negativity constraint.

(iv) Setting $5x_1 + 2x_2 = 4$ active gives

$$L(\mathbf{x}, \lambda) = x_1^2 + 4x_2^2 - 2x_1 + 8x_2 + \lambda(5x_1 + 2x_2 - 4)$$

with necessary conditions:

$$\begin{aligned} \frac{\partial L}{\partial x_1} &= 2x_1 - 2 + 5\lambda = 0 \\ \frac{\partial L}{\partial x_2} &= 8x_2 + 8 + 2\lambda = 0 \\ \frac{\partial L}{\partial \lambda} &= 5x_1 + 2x_2 - 4 = 0. \end{aligned}$$

Solving gives $\lambda = -\frac{1}{13}$, $x_1 = 1.92$ and $x_2 = -0.98 < 0$, which also violates a non-negativity constraint.

(v) Now setting $x_2 = 0$ in addition to $5x_1 + 2x_2 - 4 = 0$ results in an additional term $-\lambda_2 x_2$ in L giving

$$\frac{\partial L}{\partial x_1} = 2x_1 - 2 + 5\lambda_1 = 0 \text{ and } \frac{\partial L}{\partial \lambda} = 8x_2 + 8 + 2\lambda_1 - \lambda_2 = 0.$$

Solving together with the equalities gives $x_1 = \frac{4}{5}$; $x_2 = 0$ which satisfies all the constraints and with $\lambda_1 = \frac{2}{25} > 0$ and $\lambda_2 = \frac{4}{25} > 0$.

(vi) Finally setting $x_1 = 0$ and $5x_1 + 2x_2 - 4 = 0$ leads to the solution $x_1 = 0$, $x_2 = 2$ but with both associated λ's negative.

Thus the solution corresponds to case (v) i.e.:

$$x_1 = \tfrac{4}{5},\ x_2 = 0 \text{ with } f^* = -\tfrac{24}{25}.$$

Problem 5.3.6.4

Minimize $f(\mathbf{x}) = \frac{1}{2}\mathbf{x}^T\mathbf{A}\mathbf{x} + \mathbf{b}^T\mathbf{x}$, $\mathbf{A} = \begin{bmatrix} 1 & -1 \\ -1 & 2 \end{bmatrix}$, $\mathbf{b} = \begin{bmatrix} -2 \\ 1 \end{bmatrix}$ such that

$$\begin{aligned} x_1,\ x_2 &\geq 0 \\ x_1 + x_2 &\leq 3 \\ 2x_1 - x_2 &\leq 4. \end{aligned}$$

Solution

First determine the unconstrained minimum:

$$\mathbf{x}^0 = -\mathbf{A}^{-1}\mathbf{b} = -\begin{bmatrix} 2 & 1 \\ 1 & 1 \end{bmatrix}\begin{bmatrix} -2 \\ 1 \end{bmatrix} = \begin{bmatrix} 3 \\ 1 \end{bmatrix}$$

$x_1,\ x_2 > 0$, $x_1 + x_2 = 3 + 1 = 4 > 3$ (constraint violation), and $2x_1 - x_2 = 5 > 4$ (constraint violation). Now consider the violated constraints separately active.

Firstly for $x_1 + x_2 - 3 = 0$ the Lagrangian is

$$L(\mathbf{x}, \lambda) = \tfrac{1}{2}\mathbf{x}^T\mathbf{A}\mathbf{x} + \mathbf{b}^T\mathbf{x} + \lambda(x_1 + x_2 - 3)$$

with necessary conditions:

$$\begin{bmatrix} \mathbf{A} & & 1 \\ & & 1 \\ 1 & 1 & 0 \end{bmatrix}\begin{bmatrix} \mathbf{x} \\ \lambda \end{bmatrix} = \begin{bmatrix} 2 \\ -1 \\ 3 \end{bmatrix}, \text{ i.e. } \begin{bmatrix} 1 & -1 & 1 \\ -1 & 2 & 1 \\ 1 & 1 & 0 \end{bmatrix}\begin{bmatrix} x_1 \\ x_2 \\ \lambda \end{bmatrix} = \begin{bmatrix} 2 \\ -1 \\ 3 \end{bmatrix}.$$

The solution of which is given by Cramer's rule:

$$x_1 = \tfrac{D_1}{D} = \tfrac{-12}{-5} = 2.4,\quad x_2 = \tfrac{D_2}{D} = \tfrac{-3}{-5} = 0.6 \text{ and } \lambda = \tfrac{D_3}{D} = \tfrac{-1}{-5} = 0.2.$$

This solution, however violates the last inequality constraint.

Similarly, setting $2x_1 - x_2 - 4 = 0$ gives the solution $\begin{bmatrix} x_1 \\ x_2 \\ \lambda \end{bmatrix} = \begin{bmatrix} 2.4 \\ 0.8 \\ 0.2 \end{bmatrix}$, which violates the first constraint.

Finally try both constraints simultaneously active. This results in a system of four linear equations in four unknowns, the solution of which (do self) is $x_1 = 2\frac{1}{3}$, $x_2 = \frac{2}{3}$ and $\lambda_1 = \lambda_2 = 0.11 > 0$, which satisfies the KKT conditions. Thus $\mathbf{x}^* = [2\frac{1}{3}, \frac{2}{3}]^T$ with $f(\mathbf{x}^*) = 1.8$.

5.3.7 Application of the gradient projection method

Problem 5.3.7.1

Apply the gradient projection method to the following problem:

minimize $f(\mathbf{x}) = x_1^2 + x_2^2 + x_3^2 - 2x_1$

such that $2x_1 + x_2 + x_3 = 7$; $x_1 + x_2 + 2x_3 = 6$ and given initial starting point $\mathbf{x}^0 = [2, 2, 1]^T$. Perform the minimization for the first projected search direction only.

Solution

$$\mathbf{P} = \left(\mathbf{I} - \mathbf{A}^T(\mathbf{A}\mathbf{A}^T)^{-1}\mathbf{A}\right),\quad \mathbf{A} = \begin{bmatrix} 2 & 1 & 1 \\ 1 & 1 & 2 \end{bmatrix},$$

$$\mathbf{A}\mathbf{A}^T = \begin{bmatrix} 6 & 5 \\ 5 & 6 \end{bmatrix},\quad \left(\mathbf{A}\mathbf{A}^T\right)^{-1} = \tfrac{1}{11}\begin{bmatrix} 6 & -5 \\ -5 & 6 \end{bmatrix},$$

$$\left(\mathbf{A}\mathbf{A}^T\right)^{-1}\mathbf{A} = \tfrac{1}{11}\begin{bmatrix} 6 & -5 \\ -5 & 6 \end{bmatrix}\begin{bmatrix} 2 & 1 & 1 \\ 1 & 1 & 2 \end{bmatrix} = \tfrac{1}{11}\begin{bmatrix} 7 & 1 & -4 \\ -4 & 1 & 7 \end{bmatrix},$$

$$\mathbf{A}^T\left(\mathbf{A}\mathbf{A}^T\right)^{-1}\mathbf{A} = \tfrac{1}{11}\begin{bmatrix} 2 & 1 \\ 1 & 1 \\ 1 & 2 \end{bmatrix}\begin{bmatrix} 7 & 1 & -4 \\ -4 & 1 & 7 \end{bmatrix} = \tfrac{1}{11}\begin{bmatrix} 10 & 3 & -1 \\ 3 & 2 & 3 \\ -1 & 3 & 10 \end{bmatrix}.$$

It now follows that

$$\mathbf{P} = \tfrac{1}{11}\begin{bmatrix} 1 & -3 & 1 \\ -3 & 9 & -3 \\ 1 & -3 & 1 \end{bmatrix} \text{ and } \nabla f(\mathbf{x}^0) = \begin{bmatrix} 2 \\ 4 \\ 2 \end{bmatrix}.$$

This gives the projected steepest descent direction as $-\mathbf{P}\nabla f = -\frac{1}{11}\begin{bmatrix} -8 \\ 24 \\ -8 \end{bmatrix}$

and therefore select the descent search direction as $\mathbf{u}^1 = [1, -3, 1]^T$.

Along the line through $\mathbf{x}^0$ in the direction $\mathbf{u}^1$, $\mathbf{x}$ is given by

$$\begin{aligned} \mathbf{x} &= \mathbf{x}^0 + \lambda \begin{bmatrix} 1 \\ -3 \\ 1 \end{bmatrix} = \begin{bmatrix} 2+\lambda \\ 2-3\lambda \\ 1+\lambda \end{bmatrix} \text{ and} \\ f(\mathbf{x}^0 + \lambda\mathbf{u}) &= F(\lambda) = (2+\lambda)^2 + (2-3\lambda)^2 + (1+\lambda)^2 - 2(2+\lambda). \end{aligned}$$

For a minimum $\frac{dF}{d\lambda} = 2(2+\lambda) - 6(2-3\lambda) + 2(1+\lambda) - 2 = 0$, which gives $\lambda_1 = \frac{4}{11}$. Thus the next iterate is

$$\mathbf{x}^1 = (2 + \tfrac{4}{11}; 2 - \tfrac{12}{11}; 1 + \tfrac{4}{11})^T = (2\tfrac{4}{11}; \tfrac{10}{11}; 1\tfrac{4}{11})^T.$$

Problem 5.3.7.2

Apply the gradient projection method to the minimization of

$f(\mathbf{x}) = x_1^2 + x_2^2 - 2x_1 - 4x_2$ such that $h(\mathbf{x}) = x_1 + 4x_2 - 5 = 0$

with starting point $\mathbf{x}^0 = [1; 1]^T$.

Solution

The projection matrix is $\mathbf{P} = \mathbf{I} - \mathbf{A}^T(\mathbf{A}\mathbf{A}^T)^{-1}\mathbf{A}$.

Here $\mathbf{A} = [1 \;\; 4]$ and therefore

$$\mathbf{P} = \begin{bmatrix} 1 & 0 \\ 0 & 1 \end{bmatrix} - \begin{bmatrix} 1 \\ 4 \end{bmatrix} \left([1 \;\; 4] \begin{bmatrix} 1 \\ 4 \end{bmatrix} \right)^{-1} [1 \;\; 4] = \frac{1}{17}\begin{bmatrix} 16 & -4 \\ -4 & 1 \end{bmatrix}.$$

With $\nabla f(\mathbf{x}^0) = \begin{bmatrix} 2x_1^0 - 2 \\ 2x_2^0 - 4 \end{bmatrix} = \begin{bmatrix} 0 \\ -2 \end{bmatrix}$ the search direction is

$$\mathbf{u}^1 = -\mathbf{P}\nabla f(\mathbf{x}^0) = -\frac{1}{17}\begin{bmatrix} 16 & -4 \\ -4 & 1 \end{bmatrix}\begin{bmatrix} 0 \\ -2 \end{bmatrix} = -\frac{1}{17}\begin{bmatrix} 8 \\ -2 \end{bmatrix}$$

or more conveniently $\mathbf{u}^1 = \begin{bmatrix} -4 \\ 1 \end{bmatrix}$.

Thus $\mathbf{x} = \mathbf{x}^0 + \lambda\mathbf{u}^1 = \begin{bmatrix} 1-4\lambda \\ 1+\lambda \end{bmatrix}$ for the line search. Along the search line

$$F(\lambda) = (1-4\lambda)^2 + (1+\lambda)^2 - 2(1-4\lambda) + 4(1+\lambda) = 17\lambda^2 - 2\lambda - 4$$

with minimum occurring where $\frac{dF}{d\lambda} = 34\lambda - 2 = 0$, giving $\lambda_1 = \frac{1}{17}$.

Thus $\mathbf{x}^1 = \mathbf{x}^0 + \frac{1}{17}\mathbf{u}^1 = \begin{bmatrix} 1 \\ 1 \end{bmatrix} + \frac{1}{17}\begin{bmatrix} -4 \\ 1 \end{bmatrix} = \begin{bmatrix} 0.7647 \\ 1.0588 \end{bmatrix}$.

Next, compute $\mathbf{u}^2 = -\mathbf{P}\nabla f(\mathbf{x}^1) = -\frac{1}{17}\begin{bmatrix} 16 & -4 \\ -4 & 1 \end{bmatrix}\begin{bmatrix} 0.4706 \\ 1.8824 \end{bmatrix} = \begin{bmatrix} 0.0 \\ 0.0 \end{bmatrix}$.

Since the projected gradient equals $\mathbf{0}$, the point $\mathbf{x}^1$ is the optimum $\mathbf{x}^*$.

Problem 5.3.7.3

Apply the gradient projection method to the problem:

minimize $f(\mathbf{x}) = x_1^2 + x_2^2 + x_3^2$ such that $x_1 + x_2 + x_3 = 1$

with starting point $\mathbf{x}^0 = [1, 0, 0]^T$.

Solution

$$\mathbf{P} = \mathbf{I} - \mathbf{A}^T(\mathbf{A}\mathbf{A}^T)^{-1}\mathbf{A}.$$

Here $\mathbf{A} = [1\ 1\ 1]$, $\mathbf{A}\mathbf{A}^T = 3$, $(\mathbf{A}\mathbf{A}^T)^{-1} = \frac{1}{3}$ and thus

$$\mathbf{P} = \begin{bmatrix} 1 & 0 & 0 \\ 0 & 1 & 0 \\ 0 & 0 & 1 \end{bmatrix} - \frac{1}{3}\begin{bmatrix} 1 & 1 & 1 \\ 1 & 1 & 1 \\ 1 & 1 & 1 \end{bmatrix} = \frac{1}{3}\begin{bmatrix} 2 & -1 & -1 \\ -1 & 2 & -1 \\ -1 & -1 & 2 \end{bmatrix}.$$

Since $\nabla f(\mathbf{x}) = \begin{bmatrix} 2x_1 \\ 2x_2 \\ 2x_3 \end{bmatrix}$ it follows that $\nabla f(\mathbf{x}^0) = \begin{bmatrix} 2 \\ 0 \\ 0 \end{bmatrix}$.

The projected search direction is therefore given by $\mathbf{u}^1 = -\mathbf{P}\nabla f(\mathbf{x}^0) = -\frac{1}{3}\begin{bmatrix} 2 & -1 & -1 \\ -1 & 2 & -1 \\ -1 & -1 & 2 \end{bmatrix}\begin{bmatrix} 2 \\ 0 \\ 0 \end{bmatrix} = \frac{1}{3}\begin{bmatrix} 4 \\ -2 \\ -2 \end{bmatrix} = -\frac{2}{3}\begin{bmatrix} 2 \\ -1 \\ -1 \end{bmatrix}$, or more conveniently $\mathbf{u}^1 = \begin{bmatrix} -2 \\ 1 \\ 1 \end{bmatrix}$.

The next point therefore lies along the line $\mathbf{x} = \mathbf{x}^0 + \lambda \begin{bmatrix} -2 \\ 1 \\ 1 \end{bmatrix} = \begin{bmatrix} 1-2\lambda \\ \lambda \\ \lambda \end{bmatrix}$ with

$$F(\lambda) = f(\mathbf{x}_0 + \lambda \mathbf{u}^1) = (1-2\lambda)^2 + \lambda^2 + \lambda^2.$$

The minimum occurs where $\frac{dF}{d\lambda} = -2(1-2\lambda)2 + 2\lambda + 2\lambda = 0$, giving $\lambda_1 = \frac{1}{3}$.

Thus $\mathbf{x}^1 = \left[\frac{1}{3}, \frac{1}{3}, \frac{1}{3}\right]^T$, $\nabla f(\mathbf{x}^1) = \left[\frac{2}{3}, \frac{2}{3}, \frac{2}{3}\right]^T$ and

$$\mathbf{P}\nabla f(\mathbf{x}^1) = \frac{1}{3}\begin{bmatrix} 2 & -1 & -1 \\ -1 & 2 & -1 \\ -1 & -1 & 2 \end{bmatrix} \frac{2}{3}\begin{bmatrix} 1 \\ 1 \\ 1 \end{bmatrix} = \frac{2}{9}\begin{bmatrix} 0 \\ 0 \\ 0 \end{bmatrix} = \mathbf{0}.$$

Since the projected gradient is equal to $\mathbf{0}$, the optimum point is $\mathbf{x}^* = \mathbf{x}^1$.

5.3.8 Application of the augmented Lagrangian method

Problem 5.3.8.1

Minimize $f(\mathbf{x}) = 6x_1^2 + 4x_1x_2 + 3x_2^2$ such that $h(\mathbf{x}) = x_1 + x_2 - 5 = 0$, by means of the augmented Lagrangian multiplier method.

Solution

Here $\mathcal{L}(\mathbf{x}, \lambda, \rho) = 6x_1^2 + 4x_1x_2 + 3x_2^2 + \lambda(x_1 + x_2 - 5) + \rho(x_1 + x_2 - 5)^2$ with necessary conditions for a stationary point:

$$\frac{\partial \mathcal{L}}{\partial x_1} = 0 \;\Rightarrow\; x_1(12 + 2\rho) + x_2(4 + 2\rho) = 10\rho - \lambda$$
$$\frac{\partial \mathcal{L}}{\partial x_2} = 0 \;\Rightarrow\; x_1(4 + 2\rho) + x_2(6 + 2\rho) = 10\rho - \lambda.$$

With $\rho = 1$ the above become $14x_1 + 6x_2 = 10 - \lambda = 6x_1 + 8x_2$ and it follows that

$$x_2 = 4x_1 \text{ and } x_1 = \frac{10-\lambda}{38}.$$

The iterations now proceed as follows.

Iteration 1: $\lambda^1 = 0$

$$x_1 = \tfrac{10}{38} = 0.2632, \quad x_2 = 1.0526$$

$$h = x_1 + x_2 - 5 = 0.2632 + 1.0526 - 5 = -3.6842$$

and for the next iteration

$$\lambda^2 = \lambda^1 + 2\rho h = 0 + (2)(1)(-3.6842) = -7.3684.$$

Iteration 2:

$$x_1 = \tfrac{10-\lambda}{38} = \tfrac{10+7.3684}{38} = 0.4571$$

and it follows that $x_2 = 1.8283$ and thus $h = 0.4571 + 1.8283 - 5 = -2.7146$ with the next multiplier value given by

$$\lambda^3 = \lambda^2 + (2)(1)h = -7.3684 + 2(-2.7146) = -12.7978.$$

Iteration 3:

Now increase ρ to 10; then $x_1 = \frac{100-\lambda}{128}$ and $x_2 = 4x_1$.

Thus $x_1 = \frac{100+12.7978}{128} = 0.8812$ and $x_2 = 3.5249$ with

$$\lambda^4 = -12.7978 + 2(10)(-0.5939) = -24.675.$$

Iteration 4:

$$x_1 = \tfrac{100+24.675}{128} = 0.9740, \; x_2 = 3.8961 \text{ with } \lambda^5 = -27.27.$$

Iteration 5: Iteration 5 gives $x_1 = 0.9943$, $x_2 = 3.9772$ with $\lambda^6 = -27.84$ and rapid convergence is obtained to the solution $\mathbf{x}^* = [1, 4]^T$.

5.3.9 Application of the sequential quadratic programming method

Problem 5.3.9.1

Minimize $f(\mathbf{x}) = 2x_1^2 + 2x_2^2 - 2x_1x_2 - 4x_1 - 6x_2$, such that $h(\mathbf{x}) = 2x_1^2 - x_2 = 0$ by means of the SQP method and starting point $\mathbf{x}^0 = [0, 1]^T$.

Solution

Here $L(\mathbf{x}, \lambda) = f(\mathbf{x}) + \lambda h(\mathbf{x})$ and the necessary conditions for a stationary point are

$$\begin{aligned}\frac{\partial L}{\partial x_1} &= 4x_1 - 2x_2 - 4 + \lambda 4x_1 = 0\\ \frac{\partial L}{\partial x_2} &= 4x_2 - 2x_1 - 6 - \lambda = 0\\ h(\mathbf{x}) &= 2x_1^2 - x_2 = 0.\end{aligned}$$

This system of non-linear equations may be written in vector form as $\mathbf{c}(\mathbf{X}) = \mathbf{0}$, where $\mathbf{X} = [x_1, x_2, \lambda]^T$. If an approximate solution $(\mathbf{x}^k, \lambda^k)$ to this system is available a possible improvement may be obtained by solving the linearized system via Newton's method:

$$\left[\frac{\partial \mathbf{c}}{\partial \mathbf{x}}\right]^T \begin{bmatrix} \mathbf{x} - \mathbf{x}^k \\ \lambda - \lambda^k \end{bmatrix} = -\mathbf{c}(\mathbf{x}^k, \lambda^k),$$

where $\frac{\partial \mathbf{c}}{\partial \mathbf{x}}$ is the Jacobian of the system $\mathbf{c}$.

In detail this linear system becomes

$$\begin{bmatrix} 4 & -2 & 4x_1 \\ -2 & 4 & -1 \\ 4x_1 & -1 & 0 \end{bmatrix} \begin{bmatrix} \mathbf{x} - \mathbf{x}^k \\ \\ \lambda - \lambda^k \end{bmatrix} = - \begin{bmatrix} c_1 \\ c_2 \\ c_3 \end{bmatrix}$$

and for the approximation $\mathbf{x}^0 = [0, 1]^T$ and $\lambda^0 = 0$, the system may be written as

$$\begin{bmatrix} 4 & -2 & 0 \\ -2 & 4 & -1 \\ 0 & -1 & 0 \end{bmatrix} \begin{bmatrix} s_1 \\ s_2 \\ \Delta\lambda \end{bmatrix} = - \begin{bmatrix} -6 \\ -2 \\ -1 \end{bmatrix}$$

which has the solution $s_1 = 1$, $s_2 = -1$ and $\Delta\lambda = -8$. Thus after the first iteration $x_1^1 = 0 + 1 = 1$, $x_2^1 = 1 - 1 = 0$ and $\lambda^1 = 0 - 8 = -8$.

The above Newton step is equivalent to the solution of the following quadratic programming (QP) problem set up at $\mathbf{x}^0 = [0, 1]^T$ with $\lambda^0 = 0$ (see Section 3.5.3):

$$\text{minimize } F(\mathbf{s}) = \tfrac{1}{2}\mathbf{s}^T \begin{bmatrix} 4 & -2 \\ -2 & 4 \end{bmatrix} \mathbf{s} + [-6\ -2]\mathbf{s} - 4 \text{ such that } s_2 + 1 = 0.$$

Setting $s_2 = -1$ gives $F(\mathbf{s}) = F(s_1) = 2s_1^2 - 4s_1 - 2$ with minimum at $s_1 = 1$. Thus after the first iteration $x_1^1 = s_1 = 1$ and $x_2^1 = s_2 + 1 = 0$, giving $\mathbf{x}^1 = [1, 0]^T$, which corresponds to the Newton step solution.

Here, in order to simplify the computation, λ is not updated. Continue by setting up the next QP: with $f(\mathbf{x}^1) = -2$; $\boldsymbol{\nabla} f(\mathbf{x}^1) = [0, -8]^T$; $h(\mathbf{x}^1) = 2$; $\boldsymbol{\nabla} h(\mathbf{x}^1) = [4, -1]^T$ the QP becomes:

$$\begin{aligned} \text{minimize}\, F(\mathbf{s}) &= \frac{1}{2}\mathbf{s}^T \begin{bmatrix} 4 & -2 \\ -2 & 4 \end{bmatrix} \mathbf{s} + [0, -8]\mathbf{s} - 2 \\ &= 2s_1^2 - 2s_1s_2 + 2s_2^2 - 8s_2 - 2 \end{aligned}$$

with constraint

$$\begin{aligned} h(\mathbf{s}) &= [4, -1] \begin{bmatrix} s_1 \\ s_2 \end{bmatrix} + 2 = 0 \\ &= 4s_1 - s_2 + 2 = 0. \end{aligned}$$

Substituting $s_2 = 4s_1 + 2$ in $F(\mathbf{s})$ results in the unconstrained minimization of the single variable function:

$$F(s_1) = 26s_1^2 - 4s_1 - 10.$$

Setting $\frac{dF}{ds_1} = 52s_1 - 4 = 0$, gives $s_1 = 0.07692$ and $s_2 = 2.30769$. Thus $\mathbf{x}^2 = [1 + s_1, 0 + s_2] = [1.07692, 2.30769]^T$. Continuing in this manner yields $\mathbf{x}^3 = [1.06854, 2.2834]^T$ and $\mathbf{x}^4 = [1.06914, 2.28575]^T$, which represents rapid convergence to the exact optimum $\mathbf{x}^* = [1.06904, 2.28569]^T$, with $f(\mathbf{x}^*) = -10.1428$. Substituting $\mathbf{x}^*$ into the necessary conditions for a stationary point of L, gives the value of $\lambda^* = 1.00468$.

Chapter 6

SOME THEOREMS

6.1 Characterization of functions and minima

Theorem 6.1.1

If $f(\mathbf{x})$ is a differentiable function over the convex set $X \subseteq \mathbb{R}^n$ then $f(\mathbf{x})$ is convex over X *if and only if*

$$f(\mathbf{x}^2) \geq f(\mathbf{x}^1) + \nabla^T f(\mathbf{x}^1)(\mathbf{x}^2 - \mathbf{x}^1) \tag{6.1}$$

for all $\mathbf{x}^1$, $\mathbf{x}^2 \in X$.

Proof

If $f(\mathbf{x})$ is convex over X then by the definition (1.9), for all $\mathbf{x}^1, \mathbf{x}^2 \in X$ and for all $\lambda \in [0, 1]$

$$f(\lambda\mathbf{x}^2 + (1-\lambda)\mathbf{x}^1) \leq \lambda f(\mathbf{x}^2) + (1-\lambda)f(\mathbf{x}^1) \tag{6.2}$$

i.e.

$$\frac{f(\mathbf{x}^1 + \lambda(\mathbf{x}^2 - \mathbf{x}^1)) - f(\mathbf{x}^1)}{\lambda} \leq f(\mathbf{x}^2) - f(\mathbf{x}^1). \tag{6.3}$$

Taking the limit as $\lambda \to 0$ it follows that

$$\left.\frac{df(\mathbf{x}^1)}{d\lambda}\right|_{\mathbf{x}^2-\mathbf{x}^1} \leq f(\mathbf{x}^2) - f(\mathbf{x}^1). \tag{6.4}$$

The directional derivative on the left hand side may also, by (1.16) be written as

$$\left.\frac{df(\mathbf{x}^1)}{d\lambda}\right|_{\mathbf{x}^2-\mathbf{x}^1} = \nabla^T f(\mathbf{x})(\mathbf{x}^2 - \mathbf{x}^1). \tag{6.5}$$

Substituting (6.5) into (6.4) gives $f(\mathbf{x}^2) \geq f(\mathbf{x}^1) + \nabla^T f(\mathbf{x}^1)(\mathbf{x}^2 - \mathbf{x}^1)$, i.e. (6.1) is true.

Conversely, if (6.1) holds, then for $\mathbf{x} = \lambda\mathbf{x}^2 + (1-\lambda)\mathbf{x}^1 \in X$, $\lambda \in [0,1]$:

$$f(\mathbf{x}^2) \geq f(\mathbf{x}) + \nabla^T f(\mathbf{x})(\mathbf{x}^2 - \mathbf{x}) \tag{6.6}$$
$$f(\mathbf{x}^1) \geq f(\mathbf{x}) + \nabla^T f(\mathbf{x})(\mathbf{x}^1 - \mathbf{x}). \tag{6.7}$$

Multiplying (6.6) by λ and (6.7) by $(1-\lambda)$ and adding gives

$$\lambda f(\mathbf{x}^2) + (1-\lambda)f(\mathbf{x}^1) - f(\mathbf{x}) \geq \nabla^T f(\mathbf{x})(\lambda(\mathbf{x}^2 - \mathbf{x}) + (1-\lambda)(\mathbf{x}^1 - \mathbf{x})) = 0$$

since $\lambda(\mathbf{x}^2 - \mathbf{x}) + (1-\lambda)(\mathbf{x}^1 - \mathbf{x}) = \mathbf{0}$ and it follows that

$$f(\mathbf{x}) = f(\lambda\mathbf{x}^2 + (1-\lambda)\mathbf{x}^1) \leq \lambda f(\mathbf{x}^2) + (1-\lambda)f(\mathbf{x}^1),$$

i.e. $f(\mathbf{x})$ is convex and the theorem is proved. □

(Clearly if $f(\mathbf{x})$ is to be strictly convex the strict inequality > applies in (6.1).)

Theorem 6.1.2

If $f(\mathbf{x}) \in C^2$ over an open convex set $X \subseteq \mathbb{R}^n$, then $f(\mathbf{x})$ is convex *if and only if* $\mathbf{H}(\mathbf{x})$ is positive semi-definite for all $\mathbf{x} \in X$.

Proof

If $\mathbf{H}(\mathbf{x})$ is positive semi-definite for all $\mathbf{x} \in X$, then for all $\mathbf{x}^1$, $\mathbf{x}^2$ in X it follows by the Taylor expansion (1.21) that

$$\begin{aligned} f(\mathbf{x}^2) = f(\mathbf{x}^1 + (\mathbf{x}^2 - \mathbf{x}^1)) &= f(\mathbf{x}^1) + \nabla^T f(\mathbf{x}^1)(\mathbf{x}^2 - \mathbf{x}^1) \\ &+ \tfrac{1}{2}(\mathbf{x}^2 - \mathbf{x}^1)^T \mathbf{H}(\bar{\mathbf{x}})(\mathbf{x}^2 - \mathbf{x}^1) \end{aligned} \tag{6.8}$$

where $\bar{\mathbf{x}} = \mathbf{x}^1 + \theta(\mathbf{x}^2 - \mathbf{x}^1))$, $\theta \in [0,1]$. Since $\mathbf{H}(\mathbf{x})$ is positive semi-definite it follows directly from (6.8) that $f(\mathbf{x}^2) \geq f(\mathbf{x}^1) + \nabla^T f(\mathbf{x}^1)(\mathbf{x}^2 - \mathbf{x}^1)$ and therefore by Theorem 6.1.1 $f(\mathbf{x})$ is convex.

Conversely if $f(\mathbf{x})$ is convex, then from (6.1) for all $\mathbf{x}^1$, $\mathbf{x}^2$ in X

$$f(\mathbf{x}^2) \geq f(\mathbf{x}^1) + \boldsymbol{\nabla}^T f(\mathbf{x}^1)(\mathbf{x}^2 - \mathbf{x}^1). \tag{6.9}$$

Also the Taylor expansion (6.8) above applies, and comparison of (6.8) and (6.9) implies that

$$\frac{1}{2}(\mathbf{x}^2 - \mathbf{x}^1)^T \mathbf{H}(\bar{\mathbf{x}})(\mathbf{x}^2 - \mathbf{x}^1) \geq 0. \tag{6.10}$$

Clearly since (6.10) must apply for all $\mathbf{x}^1$, $\mathbf{x}^2$ in X and since $\bar{\mathbf{x}}$ is assumed to vary continuously with $\mathbf{x}^1$, $\mathbf{x}^2$, (6.10) must be true for any $\bar{\mathbf{x}}$ in X, i.e. $\mathbf{H}(\mathbf{x})$ is positive semi-definite for all $\mathbf{x} \in X$, which concludes the proof. □

Theorem 6.1.3

If $f(\mathbf{x}) \in C^2$ over an open convex set $X \subseteq \mathbb{R}^n$, then if the Hessian matrix $\mathbf{H}(\mathbf{x})$ is positive-definite for all $\mathbf{x} \in X$, then $f(\mathbf{x})$ is *strictly* convex over X.

Proof

For any $\mathbf{x}^1$, $\mathbf{x}^2$ in X it follows by the Taylor expansion (1.21) that

$$\begin{aligned} f(\mathbf{x}^2) &= f(\mathbf{x}^1 + (\mathbf{x}^2 - \mathbf{x}^1)) \\ &= f(\mathbf{x}^1) + \boldsymbol{\nabla}^T f(\mathbf{x}^1)(\mathbf{x}^2 - \mathbf{x}^1) + \frac{1}{2}(\mathbf{x}^2 - \mathbf{x}^1)^T \mathbf{H}(\bar{\mathbf{x}})(\mathbf{x}^2 - \mathbf{x}^1) \end{aligned}$$

where $\bar{\mathbf{x}} = \mathbf{x}^1 + \theta(\mathbf{x}^2 - \mathbf{x}^1)$, $\theta \in [0, 1]$. Since $\mathbf{H}(\mathbf{x})$ is positive-definite it follows directly that $f(\mathbf{x}^2) > f(\mathbf{x}^1) + \boldsymbol{\nabla}^T f(\mathbf{x}^1)(\mathbf{x}^2 - \mathbf{x}^1)$ and therefore by Theorem 6.1.1 $f(\mathbf{x})$ is strictly convex and the theorem is proved. □

Theorem 6.1.4

If $f(\mathbf{x}) \in C^2$ over the convex set $X \subseteq \mathbb{R}^n$ and if $f(\mathbf{x})$ is convex over X, then any interior local minimum of $f(\mathbf{x})$ is a global minimum.

Proof

If $f(\mathbf{x})$ is convex then by (6.1):

$$f(\mathbf{x}^2) \geq f(\mathbf{x}^1) + \boldsymbol{\nabla}^T f(\mathbf{x}^1)(\mathbf{x}^2 - \mathbf{x}^1)$$

for all $\mathbf{x}^1, \mathbf{x}^2 \in X$. In particular for any point $\mathbf{x}^2 = \mathbf{x}$ and in particular $\mathbf{x}^1 = \mathbf{x}^*$ an interior local minimum, it follows that

$$f(\mathbf{x}) \geq f(\mathbf{x}^*) + \boldsymbol{\nabla}^T f(\mathbf{x}^*)(\mathbf{x} - \mathbf{x}^*).$$

Since the necessary condition $\boldsymbol{\nabla}^T f(\mathbf{x}^*) = \mathbf{0}$ applies at $\mathbf{x}^*$, the above reduces to $f(\mathbf{x}) \geq f(\mathbf{x}^*)$ and therefore $\mathbf{x}^*$ is a global minimum. □

Theorem 6.1.5

More generally, let $f(\mathbf{x})$ be strictly convex on the convex set X, but $f(\mathbf{x})$ not necessarily $\in C^2$, then a strict local minimum is the global minimum.

Proof

Let $\mathbf{x}^0$ be the strict local minimum in a δ-neighbourhood and $\mathbf{x}^*$ the global minimum, $\mathbf{x}^0, \mathbf{x}^* \in X$ and assume that $\mathbf{x}^0 \neq \mathbf{x}^*$. Then there exists an ε, $0 < \varepsilon < \delta$, with $f(\mathbf{x}) > f(\mathbf{x}^0)$ for all $\mathbf{x}$ such that $\|\mathbf{x} - \mathbf{x}^0\| < \varepsilon$.

A convex combination of $\mathbf{x}^0$ and $\mathbf{x}^*$ is given by

$$\widehat{\mathbf{x}} = \lambda \mathbf{x}^* + (1 - \lambda)\mathbf{x}^0.$$

Note that if $\lambda \rightarrow 0$ then $\widehat{\mathbf{x}} \rightarrow \mathbf{x}^0$.

As $f(\mathbf{x})$ is strictly convex it follows that

$$f(\widehat{\mathbf{x}}) < \lambda f(\mathbf{x}^*) + (1 - \lambda) f(\mathbf{x}^0) \leq f(\mathbf{x}^0) \text{ for all } \lambda \in (0, 1).$$

In particular this holds for λ arbitrarily small and hence for λ such that $\|\widehat{\mathbf{x}}(\lambda) - \mathbf{x}^0\| < \varepsilon$. But as $\mathbf{x}^0$ is the strict local minimum

$$f(\widehat{\mathbf{x}}) > f(\mathbf{x}^0) \text{ for all } \widehat{\mathbf{x}} \text{ with } \|\widehat{\mathbf{x}} - \mathbf{x}^0\| < \varepsilon.$$

This contradicts the fact that

$$f(\widehat{\mathbf{x}}) < f(\mathbf{x}^0)$$

that followed from the convexity and the assumption that $\mathbf{x}^0 \neq \mathbf{x}^*$. It follows therefore that $\mathbf{x}^0 \equiv \mathbf{x}^*$ which completes the proof. □

6.2 Equality constrained problem

Theorem 6.2.1

In problem (3.5), let f and $h_j \in C^1$ and assume that the Jacobian matrix $\left[\frac{\partial \mathbf{h}(\mathbf{x}^*)}{\partial \mathbf{x}}\right]$ is of rank r. Then the *necessary conditions* for a bounded internal local minimum $\mathbf{x}^*$ of the equality constrained problem (3.5) is that $\mathbf{x}^*$ must coincide with the stationary point $(\mathbf{x}^*, \boldsymbol{\lambda}^*)$ of the Lagrange function L, i.e. that there exits a $\boldsymbol{\lambda}^*$ such that

$$\frac{\partial L}{\partial x_i}(\mathbf{x}^*, \boldsymbol{\lambda}^*) = 0, \ i = 1, 2, \ldots, n; \ \frac{\partial L}{\partial \lambda_j}(\mathbf{x}^*, \boldsymbol{\lambda}^*) = 0, \ j = 1, 2, \ldots, r.$$

Note: Here the elements of the Jacobian matrix are taken as $\left[\frac{\partial \mathbf{h}}{\partial \mathbf{x}}\right]_{ij} = \frac{\partial h_j}{\partial x_i}$, i.e. $\left[\frac{\partial \mathbf{h}}{\partial \mathbf{x}}\right] = [\boldsymbol{\nabla} h_1, \boldsymbol{\nabla} h_2, \ldots, \boldsymbol{\nabla} h_r]$.

Proof

Since f and $h_j \in C^1$, it follows that for a bounded internal local minimum at $\mathbf{x} = \mathbf{x}^*$, that

$$df = \boldsymbol{\nabla}^T f(\mathbf{x}^*) d\mathbf{x} = 0 \tag{6.11}$$

since $df \geq 0$ for $d\mathbf{x}$ and for $-d\mathbf{x}$, for all perturbations $d\mathbf{x}$ which are consistent with the constraints, i.e. for all $d\mathbf{x}$ such that

$$dh_j = \boldsymbol{\nabla}^T h_j(\mathbf{x}^*) d\mathbf{x} = 0, \ j = 1, 2, \ldots, r. \tag{6.12}$$

Consider the Lagrange function

$$L(\mathbf{x}, \boldsymbol{\lambda}) = f(\mathbf{x}) + \sum_{j=1}^{r} \lambda_j h_j(\mathbf{x}).$$

The differential of L is given by

$$dL = df + \sum_{j=1}^{r} \lambda_j dh_j$$

and it follows from (6.11) and (6.12) that at $\mathbf{x}^*$

$$dL = \frac{\partial L}{\partial x_1}dx_1 + \frac{\partial L}{\partial x_2}dx_2 + \cdots + \frac{\partial L}{\partial x_n}dx_n = 0 \tag{6.13}$$

for all $d\mathbf{x}$ such that $\mathbf{h}(\mathbf{x}) = \mathbf{0}$, i.e. $\mathbf{h}(\mathbf{x}^* + d\mathbf{x}) = \mathbf{0}$.

Choose Lagrange multipliers λ_j, $j = 1, \ldots, r$ such that at $\mathbf{x}^*$

$$\frac{\partial L}{\partial x_j}(\mathbf{x}^*, \boldsymbol{\lambda}) = \frac{\partial f}{\partial x_j}(\mathbf{x}^*) + \left[\frac{\partial \mathbf{h}}{\partial x_j}(\mathbf{x}^*)\right]^T \boldsymbol{\lambda} = 0, \quad j = 1, 2, \ldots, r. \tag{6.14}$$

The solution of this system provides the vector $\boldsymbol{\lambda}^*$. Here the r variables, x_j, $j = 1, 2, \ldots, r$, may be any appropriate set of r variables from the set x_i, $i = 1, 2, \ldots, n$. A unique solution for $\boldsymbol{\lambda}^*$ exists as it is assumed that $\left[\frac{\partial \mathbf{h}(\mathbf{x}^*)}{\partial \mathbf{x}}\right]$ is of rank r. It follows that (6.13) can now be written as

$$dL = \frac{\partial L}{\partial x_{r+1}}(\mathbf{x}^*, \boldsymbol{\lambda}^*)dx_{r+1} + \cdots + \frac{\partial L}{\partial x_n}(\mathbf{x}^*, \boldsymbol{\lambda}^*)dx_n = 0. \tag{6.15}$$

Again consider the constraints $h_j(\mathbf{x}) = 0$, $j = 1, 2, \ldots, r$. If these equations are considered as a system of r equations in the unknowns x_1, x_2, ..., x_r, these dependent unknowns can be solved for in terms of x_{r+1}, ..., x_n. Hence the latter $n - r$ variables are the independent variables. For any choice of these independent variables, the other dependent variables x_1, ..., x_r, are determined by solving $\mathbf{h}(\mathbf{x}) = [h_1(\mathbf{x}), h_2(\mathbf{x}), \ldots, h_r(\mathbf{x})]^T = \mathbf{0}$. In particular x_{r+1} to x_n may by varied one by one at $\mathbf{x}^*$, and it follows from (6.15) that

$$\frac{\partial L}{\partial x_j}(\mathbf{x}^*, \boldsymbol{\lambda}^*) = 0, \quad j = r + 1, \ldots, n$$

and, together with (6.14) and the constraints $\mathbf{h}(\mathbf{x}) = \mathbf{0}$, it follows that the necessary conditions for a bounded internal local minimum can be written as

$$\begin{aligned} \frac{\partial L}{\partial x_i}(\mathbf{x}^*, \boldsymbol{\lambda}^*) &= 0, \quad i = 1, 2, \ldots, n \\ \frac{\partial L}{\partial \lambda_j}(\mathbf{x}^*, \boldsymbol{\lambda}^*) &= 0, \quad j = 1, 2, \ldots, r \end{aligned} \tag{6.16}$$

or

$$\nabla_{\mathbf{x}} L(\mathbf{x}^*, \boldsymbol{\lambda}^*) = \mathbf{0} \text{ and } \nabla_{\boldsymbol{\lambda}} L(\mathbf{x}^*, \boldsymbol{\lambda}^*) = \mathbf{0}.$$

□

Note that (6.16) provides $n+r$ equations in the $n+r$ unknowns x_1^*, x_2^*, ..., x_n^*, λ_1^*, ..., λ_r^*. The solutions of these, in general non-linear, equations will be candidate solutions for problem (3.5).

As a general rule, if possible, it is advantageous to solve explicitly for any of the variables in the equality constraints in terms of the others, so as to reduce the number of variables and constraints in (3.5).

Note on the existence of $\boldsymbol{\lambda}^*$

Up to this point it has been assumed that $\boldsymbol{\lambda}^*$ does indeed exist. In Theorem 6.2.1 it is assumed that the equations

$$\frac{\partial L}{\partial x_j}(\mathbf{x}^*, \boldsymbol{\lambda}) = 0$$

apply for a certain appropriate set of r variables from the set x_i, $i = 1, 2, \ldots, n$. Thus $\boldsymbol{\lambda}$ may be solved for to find $\boldsymbol{\lambda}^*$, via the linear system

$$\frac{\partial f}{\partial x_j}(\mathbf{x}^*) + \left[\frac{\partial \mathbf{h}}{\partial x_j}\right]^T \boldsymbol{\lambda} = 0, \quad j = 1, 2, \ldots, r.$$

This system can be solved if there exists a $r \times r$ submatrix $H_r \subset \left[\frac{\partial \mathbf{h}}{\partial \mathbf{x}}\right]^*$, evaluated at $\mathbf{x}^*$, such that H_r is non-singular. This is the same as requiring that $\left[\frac{\partial \mathbf{h}}{\partial \mathbf{x}}\right]^*$ be of rank r at the optimal point $\mathbf{x}^*$. This result is interesting and illuminating but, for obvious reasons, of little practical value. It does, however, emphasise the fact that it may not be assumed that multipliers will exist for every problem.

Theorem 6.2.2

If it is assumed that:

(i) $f(\mathbf{x})$ has a bounded local minimum at $\mathbf{x}^*$ (associated with this minimum there exists a $\boldsymbol{\lambda}^*$ found by solving (6.16)); and

(ii) if $\boldsymbol{\lambda}$ is chosen arbitrarily in the neighbourhood of $\boldsymbol{\lambda}^*$ then $L(\mathbf{x}, \boldsymbol{\lambda})$ has a local minimum $\mathbf{x}^0$ with respect to $\mathbf{x}$ in the neighbourhood of $\mathbf{x}^*$, then for the classical equality constrained minimization problem (3.5), the Lagrange function $L(\mathbf{x}, \boldsymbol{\lambda})$ has a saddle point at $(\mathbf{x}^*, \boldsymbol{\lambda}^*)$. Note that assumption (ii) implies that

$$\nabla_{\mathbf{x}} L(\mathbf{x}^0, \boldsymbol{\lambda}) = \mathbf{0}$$

and that a local minimum may indeed be expected if the Hessian matrix of L is positive-definite at $(\mathbf{x}^*, \boldsymbol{\lambda}^*)$.

Proof

Consider the neighbourhood of $(\mathbf{x}^*, \boldsymbol{\lambda}^*)$. As a consequence of assumption (ii), applied with $\boldsymbol{\lambda} = \boldsymbol{\lambda}^*$ $(\mathbf{x}^0 = \mathbf{x}^*)$, it follows that

$$L(\mathbf{x}, \boldsymbol{\lambda}^*) \geq L(\mathbf{x}^*, \boldsymbol{\lambda}^*) = f(\mathbf{x}^*) = L(\mathbf{x}^*, \boldsymbol{\lambda}). \tag{6.17}$$

The equality on the right holds as $\mathbf{h}(\mathbf{x}^*) = \mathbf{0}$. The relationship (6.17) shows that $L(\mathbf{x}, \boldsymbol{\lambda})$ has a degenerate saddle point at $(\mathbf{x}^*, \boldsymbol{\lambda}^*)$; for a regular saddle point it holds that

$$L(\mathbf{x}, \boldsymbol{\lambda}^*) \geq L(\mathbf{x}^*, \boldsymbol{\lambda}^*) \geq L(\mathbf{x}^*, \boldsymbol{\lambda}).$$

□

Theorem 6.2.3

If the above assumptions (i) and (ii) in Theorem 6.2.2 hold, then it follows for the bounded minimum $\mathbf{x}^*$, that

$$f(\mathbf{x}^*) = \max_{\boldsymbol{\lambda}} \left(\min_{\mathbf{x}} L(\mathbf{x}, \boldsymbol{\lambda}) \right).$$

Proof

From (ii) it is assumed that for a given $\boldsymbol{\lambda}$ the

$$\min_{\mathbf{x}} L(\mathbf{x}, \boldsymbol{\lambda}) = h(\boldsymbol{\lambda}) \text{ (the dual function)}$$

exists, and in particular for $\boldsymbol{\lambda} = \boldsymbol{\lambda}^*$

$$\min_{\mathbf{x}} L(\mathbf{x}, \boldsymbol{\lambda}^*) = h(\boldsymbol{\lambda}^*) = f(\mathbf{x}^*) \quad \text{(from (ii) } \mathbf{x}^0 = \mathbf{x}^*\text{)}.$$

Let X denote the set of points such that $\mathbf{h}(\mathbf{x}) = \mathbf{0}$ for all $\mathbf{x} \in X$. It follows that

$$\min_{\mathbf{x} \in X} L(\mathbf{x}, \boldsymbol{\lambda}) = \min_{\mathbf{x} \in X} f(\mathbf{x}) = f(\mathbf{x}^*) = h(\boldsymbol{\lambda}^*) \tag{6.18}$$

and

$$h(\boldsymbol{\lambda}) = \min_{\mathbf{x}} L(\mathbf{x}, \boldsymbol{\lambda}) \leq \min_{\mathbf{x} \in X} L(\mathbf{x}, \boldsymbol{\lambda}) = \min_{\mathbf{x} \in X} f(\mathbf{x}) = f(\mathbf{x}^*) = h(\boldsymbol{\lambda}^*). \tag{6.19}$$

Hence $h(\boldsymbol{\lambda}^*)$ is the maximum of $h(\boldsymbol{\lambda})$. Combining (6.18) and (6.19) yields that

$$f(\mathbf{x}^*) = h(\boldsymbol{\lambda}^*) = \max_{\boldsymbol{\lambda}} h(\boldsymbol{\lambda}) = \max_{\boldsymbol{\lambda}} \left(\min_{\mathbf{x}} L(\mathbf{x}, \boldsymbol{\lambda}) \right)$$

which completes the proof. □

6.3 Karush-Kuhn-Tucker theory

Theorem 6.3.1

In the problem (3.18) let f and $g_i \in C^1$, and given the existence of the Lagrange multipliers $\boldsymbol{\lambda}^*$, then the following conditions have to be satisfied at the point $\mathbf{x}^*$ that corresponds to the solution of the primal problem (3.18):

$$\begin{aligned} \frac{\partial f}{\partial x_j}(\mathbf{x}^*) + \sum_{i=1}^{m} \lambda_i^* \frac{\partial g_i}{\partial x_j}(\mathbf{x}^*) &= 0, \quad j = 1, 2, \ldots, n \\ g_i(\mathbf{x}^*) &\leq 0, \quad i = 1, 2, \ldots, m \\ \lambda_i^* g_i(\mathbf{x}^*) &= 0, \quad i = 1, 2, \ldots, m \\ \lambda_i^* &\geq 0, \quad i = 1, 2, \ldots, m \end{aligned} \tag{6.20}$$

or in more compact notation:

$$\begin{aligned} \nabla_{\mathbf{x}} L(\mathbf{x}^*, \boldsymbol{\lambda}^*) &= \mathbf{0} \\ \nabla_{\boldsymbol{\lambda}} L(\mathbf{x}^*, \boldsymbol{\lambda}^*) &\leq \mathbf{0} \\ \boldsymbol{\lambda}^{*T} \mathbf{g}(\mathbf{x}^*) &= 0 \\ \boldsymbol{\lambda}^* &\geq \mathbf{0}. \end{aligned}$$

These conditions are known as the *Karush-Kuhn-Tucker (KKT) stationary conditions.*

Proof

First convert the inequality constraints in (3.18) into equality constraints by introducing the slack variables s_i:

$$\begin{aligned} g_i(\mathbf{x}) + s_i &= 0, \quad i = 1, 2, \ldots, m \\ s_i &\geq 0. \end{aligned} \tag{6.21}$$

Define the corresponding Lagrange function

$$L(\mathbf{x}, \mathbf{s}, \boldsymbol{\lambda}) = f(\mathbf{x}) + \sum_{i=1}^{m} \lambda_i (g_i(\mathbf{x}) + s_i).$$

Assume that the solution to (3.18) with the constraints (6.21) is given by $\mathbf{x}^*$, $\mathbf{s}^*$.

Now distinguish between the two possibilities:

(i) Let $s_i^* > 0$ for all i. In this case the problem is identical to the usual minimization problem with equality constraints which is solved using Lagrange multipliers. Here there are m additional variables $s_1, s_2, \ldots, s_m$. Hence the necessary conditions for the minimum are

$$\begin{aligned} \frac{\partial L}{\partial x_j}(\mathbf{x}^*, \mathbf{s}^*, \boldsymbol{\lambda}^*) &= \frac{\partial f}{\partial x_j}(\mathbf{x}^*) + \sum_{i=1}^{m} \lambda_i^* \frac{\partial g_i}{\partial x_j}(\mathbf{x}^*) = 0, \quad j = 1, \ldots, n \\ \frac{\partial L}{\partial s_i}(\mathbf{x}^*, \mathbf{s}^*, \boldsymbol{\lambda}^*) &= \lambda_i^* = 0, \quad i = 1, \ldots, m. \end{aligned}$$

As $s_i^* > 0$, it also follows that $g_i(\mathbf{x}^*) < 0$ and with the fact that $\lambda_i^* = 0$ this yields

$$\lambda_i^* g_i(\mathbf{x}^*) = 0.$$

Consequently all the conditions of the theorem hold for the case $s_i^* > 0$ for all i.

(ii) Let $s_i^* = 0$ for $i = 1, 2, \ldots, p$ and $s_i^* > 0$ for $i = p + 1, \ldots, m$.

In this case the solution may be considered to be the solution of an equivalent minimization problem with the following equality constraints:

$$\begin{aligned} g_i(\mathbf{x}) &= 0, \quad i = 1, 2, \ldots, p \\ g_i(\mathbf{x}) + s_i &= 0, \quad i = p + 1, \ldots, m. \end{aligned}$$

Again apply the regular Lagrange theory and it follows that

$$\begin{aligned}\frac{\partial L}{\partial x_j}(\mathbf{x}^*,\mathbf{s}^*,\boldsymbol{\lambda}^*) &= \frac{\partial f}{\partial x_j}(\mathbf{x}^*)+\sum_{i=1}^{m}\lambda_i^*\frac{\partial g_i}{\partial x_j}(\mathbf{x}^*)=0,\ \ j=1,\ldots,n\\ \frac{\partial L}{\partial s_i}(\mathbf{x}^*,\mathbf{s}^*,\boldsymbol{\lambda}^*) &= \lambda_i^*=0,\ \ i=p+1,\ldots,m.\end{aligned}$$

As $g_i(\mathbf{x}^*)=0$ for $i=1,2,\ldots,p$ it follows that

$$\lambda_i^* g_i(\mathbf{x}^*)=0,\ \ i=1,2,\ldots,m.$$

Obviously

$$g_i(\mathbf{x}^*)<0,\ \ i=p+1,\ldots,m$$

and since $g_i(\mathbf{x}^*)=0$ for $i=1,2,\ldots,p$ it follows that

$$g_i(\mathbf{x}^*)\leq 0,\ \ i=1,2,\ldots,m.$$

However, no information concerning λ_i^*, $i=1,\ldots,p$ is available . This information is obtain from the following additional argument. Consider feasible changes from $\mathbf{x}^*$, $\mathbf{s}^*$ in all the variables x_1, ..., x_n, s_1, ..., s_m. Again consider m of these as dependent variables and the remaining n as independent variables. If $p\leq n$ then $s_1,s_2,\ldots,s_p$ can always be included in the set of independent variables. (Find $\boldsymbol{\lambda}^*$ by putting the partial derivatives of L at $\mathbf{x}^*$, $\mathbf{s}^*$ with respect to the dependent variables, equal to zero and solving for $\boldsymbol{\lambda}^*$.)

As $ds_i>0$ ($s_i^*=0$) must apply for feasible changes in the independent variables s_1, s_2, ..., s_p, it follows that in general for changes which are consistent with the equality constraints, that

$$df\geq 0,\qquad\qquad \text{(See Remark 2 below.)}$$

for changes involving s_1, ..., s_p. Thus if these independent variables are varied one at a time then, since all the partial derivatives of L with respect to the dependent variables must be equal to zero, that

$$df=\frac{\partial L}{\partial s_i}(\mathbf{x}^*,\boldsymbol{\lambda}^*)ds_i=\lambda_i^* ds_i\geq 0,\ \ i=1,2,\ldots,p.$$

As $ds_i > 0$ it follows that $\lambda_i \geq 0$, $i = 1, 2, \ldots, p$. Thus, since it has already been proved that $\lambda_i^* = 0$, $i = p + 1, \ldots, m$, it follows that indeed $\lambda_i^* \geq 0$, $i = 1, 2, \ldots, m$.

This completes the proof of the theorem. □

Remark 1 Obviously, if an equality constraint $h_k(\mathbf{x}) = 0$ is also prescribed explicitly, then s_k does not exist and nothing is known of the sign of λ_k^* as $\frac{\partial L}{\partial s_k}$ does not exist.

Exercise Give a brief outline of how you will obtain the necessary conditions if the equality constraints $h_j(\mathbf{x}) = 0$, $j = 1, 2, \ldots, r$ are added explicitly and L is defined by $L = f + \boldsymbol{\lambda}^T\mathbf{g} + \boldsymbol{\mu}^T\mathbf{h}$.

Remark 2 The constraints (6.21) imply that for a feasible change $ds_i > 0$ there will be a change $d\mathbf{x}$ from $\mathbf{x}^*$ and hence

$$df = \boldsymbol{\nabla}^T f d\mathbf{x} \geq 0.$$

If the condition $ds_i > 0$ does not apply, then a negative change $ds_i < 0$, equal in magnitude to the positive change considered above, would result in a corresponding change $-d\mathbf{x}$ and hence $df = \boldsymbol{\nabla}^T f(-d\mathbf{x}) \geq 0$, i.e. $df = \boldsymbol{\nabla}^T f d\mathbf{x} \leq 0$. This is only possible if $\boldsymbol{\nabla}^T f d\mathbf{x} = 0$, and consequently in this case $df = 0$.

Remark 3 It can be shown (not proved here) that the KKT stationary conditions are indeed necessary and *sufficient* conditions for a strong constrained global minimum at $\mathbf{x}^*$, if $f(\mathbf{x})$ and $g_i(\mathbf{x})$ are convex functions. This is not surprising for in the case of convex functions a local unconstrained minimum is also the global minimum.

Remark 4 Also note that if $p > n$ (see possibility (ii) of the proof) then it does not necessarily follow that $\lambda_i \geq 0$ for $i = 1, 2, \ldots, p$, i.e. the KKT conditions do not necessarily apply.

6.4 Saddle point conditions

Two drawbacks of the Kuhn-Tucker stationary conditions are that in general they only yield necessary conditions and that they apply only if $f(\mathbf{x})$ and the $g_i(\mathbf{x})$ are differentiable. These drawbacks can be removed by formulating the Karush-Kuhn-Tucker conditions in terms of the saddle point properties of $L(\mathbf{x}, \boldsymbol{\lambda})$. This is done in the next two theorems.

Theorem 6.4.1

A point $(\mathbf{x}^*, \boldsymbol{\lambda}^*)$ with $\boldsymbol{\lambda}^* \geq \mathbf{0}$ is a saddle point of the Lagrange function of the primal problem (3.18) *if and only if* the following conditions hold:

1. $\mathbf{x}^*$ minimizes $L(\mathbf{x}, \boldsymbol{\lambda}^*)$ over all $\mathbf{x}$;
2. $g_i(\mathbf{x}^*) \leq 0$, $i = 1, 2, \ldots, m$;
3. $\lambda_i^* g_i(\mathbf{x}^*) = 0$, $i = 1, 2, \ldots, m$.

Proof

If $(\mathbf{x}^*, \boldsymbol{\lambda}^*)$ is a saddle point, then $L(\mathbf{x}^*, \boldsymbol{\lambda}) \leq L(\mathbf{x}^*, \boldsymbol{\lambda}^*) \leq L(\mathbf{x}, \boldsymbol{\lambda}^*)$. First prove that if $(\mathbf{x}^*, \boldsymbol{\lambda}^*)$ is a saddle point with $\boldsymbol{\lambda}^* \geq \mathbf{0}$, then conditions 1. to 3. hold.

The right hand side of the above inequality yields directly that $\mathbf{x}^*$ minimizes $L(\mathbf{x}, \boldsymbol{\lambda}^*)$ over all $\mathbf{x}$ and the first condition is satisfied.

By expanding the left hand side:

$$f(\mathbf{x}^*) + \sum_{i=1}^{m} \lambda_i g_i(\mathbf{x}^*) \leq f(\mathbf{x}^*) + \sum_{i=1}^{m} \lambda_i^* g_i(\mathbf{x}^*)$$

and hence

$$\sum_{i=1}^{r} (\lambda_i - \lambda_i^*) g_i(\mathbf{x}^*) \leq 0 \text{ for all } \boldsymbol{\lambda} \geq \mathbf{0}.$$

Assume that $g_i(\mathbf{x}^*) > 0$. Then a contradiction is obtain for arbitrarily large λ_i, and it follows that the second condition, $g_i(\mathbf{x}^*) \leq 0$, holds.

In particular for $\boldsymbol{\lambda} = \mathbf{0}$ it follows that

$$\sum_{i=1}^{m} -\lambda_i^* g_i(\mathbf{x}^*) \leq 0 \text{ or } \sum_{i=1}^{m} \lambda_i^* g_i(\mathbf{x}^*) \geq 0.$$

But for $\boldsymbol{\lambda}^* \geq \mathbf{0}$ and $g_i(\mathbf{x}^*) \leq 0$ it follows that

$$\sum_{i=1}^{m} \lambda_i^* g_i(\mathbf{x}^*) \leq 0.$$

The only way both inequalities can be satisfied is if

$$\sum_{i=1}^{m} \lambda_i^* g_i(\mathbf{x}^*) = 0$$

and as each individual term is non positive the third condition, $\lambda_i^* g_i(\mathbf{x}^*) = 0$, follows.

Now proceed to prove the converse, i.e. that if the three conditions of the theorem hold, then $L(\mathbf{x}^*, \boldsymbol{\lambda}^*)$ has a saddle point at $(\mathbf{x}^*, \boldsymbol{\lambda}^*)$.

The first condition implies that $L(\mathbf{x}^*, \boldsymbol{\lambda}^*) \leq L(\mathbf{x}, \boldsymbol{\lambda}^*)$ which is half of the definition of a saddle point. The rest is obtained from the expansion:

$$L(\mathbf{x}^*, \boldsymbol{\lambda}) = f(\mathbf{x}^*) + \sum_{i=1}^{m} \lambda_i g_i(\mathbf{x}^*).$$

Now as $\mathbf{g}(\mathbf{x}^*) \leq \mathbf{0}$ and $\boldsymbol{\lambda} \geq \mathbf{0}$ it follows that $L(\mathbf{x}^*, \boldsymbol{\lambda}) \leq f(\mathbf{x}^*) = L(\mathbf{x}^*, \boldsymbol{\lambda}^*)$ since, from the third condition, $\sum_{i=1}^{m} \lambda_i^* g_i(\mathbf{x}^*) = 0$.

This completes the proof of the converse. □

Theorem 6.4.2

If the point $(\mathbf{x}^*, \boldsymbol{\lambda}^*)$ is a saddle point, $\boldsymbol{\lambda}^* \geq \mathbf{0}$, of the Lagrange function associated with the primal problem, then $\mathbf{x}^*$ is the solution of the primal problem.

Proof

If $(\mathbf{x}^*, \boldsymbol{\lambda}^*)$ is a saddle point the previous theorem holds and the inequality constraints are satisfied at $\mathbf{x}^*$. All that is required additionally is to show that $f(\mathbf{x}^*) \leq f(\mathbf{x})$ for all $\mathbf{x}$ such that $\mathbf{g}(\mathbf{x}) \leq \mathbf{0}$. From the definition of a saddle point it follows that

$$f(\mathbf{x}^*) + \sum_{i=1}^{m} \lambda_i^* g_i(\mathbf{x}^*) \leq f(\mathbf{x}) + \sum_{i=1}^{m} \lambda_i^* g_i(\mathbf{x})$$

for all $\mathbf{x}$ in the neighbourhood of $\mathbf{x}^*$. As a consequence of condition 3. of the previous theorem, the left hand side is $f(\mathbf{x}^*)$ and for any $\mathbf{x}$ such that $g_i(\mathbf{x}) \leq 0$, it holds that $\sum_{i=1}^{m} \lambda_i^* g_i(\mathbf{x}) \leq 0$ and hence it follows that $f(\mathbf{x}^*) \leq f(\mathbf{x})$ for all $\mathbf{x}$ such that $\mathbf{g}(\mathbf{x}) \leq \mathbf{0}$, with equality at $\mathbf{x} = \mathbf{x}^*$.

This completes the proof. □

The main advantage of these saddle point theorems is that necessary conditions are provided for solving optimization problems which are neither convex nor differentiable. Any direct search method can be used to minimize $L(\mathbf{x}, \boldsymbol{\lambda}^*)$ over all $\mathbf{x}$. Of course the problem remains that we do not have an a priori value for $\boldsymbol{\lambda}^*$. In practice it is possible to obtain estimates for $\boldsymbol{\lambda}^*$ using iterative techniques, or by solving the so called dual problem.

Theorem 6.4.3

The dual function $h(\boldsymbol{\lambda}) \leq f(\mathbf{x})$ for all $\mathbf{x}$ that satisfy the constraints $\mathbf{g}(\mathbf{x}) \leq \mathbf{0}$ for all $\boldsymbol{\lambda} \in D$. (Hence the dual function yields a lower bound for the function $f(\mathbf{x})$ with $\mathbf{g}(\mathbf{x}) \leq \mathbf{0}$.)

Proof

Let $X = \{\mathbf{x} | \mathbf{g}(\mathbf{x}) \leq \mathbf{0}\}$, then

$$\begin{aligned} h(\boldsymbol{\lambda}) &= \min_{\mathbf{x}} L(\mathbf{x}, \boldsymbol{\lambda}), \quad \boldsymbol{\lambda} \in D \\ &\leq \min_{\mathbf{x} \in X} L(\mathbf{x}, \boldsymbol{\lambda}) \\ &\leq f(\mathbf{x}) + \sum_{i=1}^{m} \lambda_i g_i(\mathbf{x}), \quad \mathbf{x} \in X \\ &\leq f(\mathbf{x}), \quad \mathbf{x} \in X, \ \boldsymbol{\lambda} \in D. \end{aligned}$$

□

The largest lower bound is attained at $\max h(\boldsymbol{\lambda})$, $\boldsymbol{\lambda} \in D$.

Theorem 6.4.4 (Duality Theorem)

The point $(\mathbf{x}^*, \boldsymbol{\lambda}^*)$ with $\boldsymbol{\lambda}^* \geq \mathbf{0}$ is a saddle point of the Lagrange function of the primal problem *if and only if*

1. $\mathbf{x}^*$ is a solution of the primal problem;

2. $\boldsymbol{\lambda}^*$ is a solution of the dual problem;

3. $f(\mathbf{x}^*) = h(\boldsymbol{\lambda}^*)$.

Proof

First assume that $L(\mathbf{x}, \boldsymbol{\lambda})$ has a saddle point at $(\mathbf{x}^*, \boldsymbol{\lambda}^*)$ with $\boldsymbol{\lambda}^* \geq \mathbf{0}$. Then from Theorem 6.4.2 it follows that $\mathbf{x}^*$ is a solution of the primal problem and 1. holds.

By definition:

$$h(\boldsymbol{\lambda}) = \min_{\mathbf{x}} L(\mathbf{x}, \boldsymbol{\lambda}).$$

From Theorem 6.4.1 it follows that $\mathbf{x}^*$ minimizes $L(\mathbf{x}, \boldsymbol{\lambda}^*)$ over all $\mathbf{x}$, thus

$$h(\boldsymbol{\lambda}^*) = f(\mathbf{x}^*) + \sum_{i=1}^{m} \lambda_i^* g_i(\mathbf{x}^*)$$

and as $\lambda_i^* g_i(\mathbf{x}^*) = 0$ it follows that

$$h(\boldsymbol{\lambda}^*) = f(\mathbf{x}^*)$$

which is condition 3.

Also by Theorem 6.4.1 $\mathbf{g}(\mathbf{x}^*) \leq \mathbf{0}$, and it has already been shown in Theorem 6.4.3 that

$$h(\boldsymbol{\lambda}) \leq f(\mathbf{x}) \text{ for all } \mathbf{x} \in \{\mathbf{x} | \mathbf{g}(\mathbf{x}) \leq \mathbf{0}\}$$

and thus in particular $h(\boldsymbol{\lambda}) \leq f(\mathbf{x}^*) = h(\boldsymbol{\lambda}^*)$. Consequently $h(\boldsymbol{\lambda}^*) = \max_{\boldsymbol{\lambda} \in D} h(\boldsymbol{\lambda})$ and condition 2. holds.

Conversely, prove that if conditions 1. to 3. hold, then $L(\mathbf{x}, \boldsymbol{\lambda})$ has a saddle point at $(\mathbf{x}^*, \boldsymbol{\lambda}^*)$, with $\boldsymbol{\lambda}^* \geq \mathbf{0}$, or equivalently that the conditions of Theorem 6.4.1 hold.

As $\mathbf{x}^*$ is a solution of the primal problem, the necessary conditions

$$\mathbf{g}(\mathbf{x}^*) \leq \mathbf{0}$$

hold, which is condition 2. of Theorem 6.4.1.

Also, as $\boldsymbol{\lambda}^*$ is a solution of the dual problem, $\boldsymbol{\lambda}^* \geq \mathbf{0}$. It is now shown that $\mathbf{x}^*$ minimizes $L(\mathbf{x}, \boldsymbol{\lambda}^*)$.

Make the contradictory assumption, i.e. that there exists a minimizer $\widehat{\mathbf{x}} \neq \mathbf{x}^*$ such that

$$L(\widehat{\mathbf{x}}, \boldsymbol{\lambda}^*) < L(\mathbf{x}^*, \boldsymbol{\lambda}^*).$$

By definition:

$$\begin{aligned} h(\boldsymbol{\lambda}^*) = L(\widehat{\mathbf{x}}, \boldsymbol{\lambda}^*) &= f(\widehat{\mathbf{x}}) + \sum_{i=1}^{m} \lambda_i^* g_i(\widehat{\mathbf{x}}) \\ &< f(\mathbf{x}^*) + \sum_{i=1}^{m} \lambda_i^* g_i(\mathbf{x}^*) \end{aligned}$$

but from condition 3.: $h(\boldsymbol{\lambda}^*) = f(\mathbf{x}^*)$ and consequently

$$\sum_{i=1}^{m} \lambda_i^* g_i(\mathbf{x}^*) > 0$$

which contradicts the fact that $\boldsymbol{\lambda}^* \geq \mathbf{0}$ and $\mathbf{g}(\mathbf{x}^*) \leq \mathbf{0}$; hence $\widehat{\mathbf{x}} = \mathbf{x}^*$ and $\mathbf{x}^*$ minimizes $L(\mathbf{x}, \boldsymbol{\lambda}^*)$ and condition 1. of Theorem 6.4.1 holds.

Also, as

$$h(\boldsymbol{\lambda}^*) = f(\mathbf{x}^*) + \sum_{i=1}^{m} \lambda_i^* g_i(\mathbf{x}^*)$$

and $h(\boldsymbol{\lambda}^*) = f(\mathbf{x}^*)$ by condition 3., it follows that:

$$\sum_{i=1}^{m} \lambda_i^* g_i(\mathbf{x}^*) = 0.$$

As each individual term is non positive the third condition of Theorem 6.4.1 holds: $\lambda_i^* g_i(\mathbf{x}^*) = 0$.

As the three conditions of Theorem 6.4.1 are satisfied it is concluded that $(\mathbf{x}^*, \boldsymbol{\lambda}^*)$ is a saddle point of $L(\mathbf{x}, \boldsymbol{\lambda})$ and the converse is proved. □

6.5 Conjugate gradient methods

Let $\mathbf{u}$ and $\mathbf{v}$ be two non-zero vectors in $\mathbb{R}^n$. Then they are mutually orthogonal if $\mathbf{u}^T\mathbf{v} = (\mathbf{u}, \mathbf{v}) = 0$. Let $\mathbf{A}$ be an $n \times n$ symmetric positive-definite matrix. Then $\mathbf{u}$ and $\mathbf{v}$ are mutually conjugate with respect to $\mathbf{A}$ if $\mathbf{u}$ and $\mathbf{Av}$ are mutually orthogonal, i.e.

$$\mathbf{u}^T\mathbf{Av} = (\mathbf{u}, \mathbf{Av}) = 0. \tag{6.22}$$

Let $\mathbf{A}$ be a square matrix. $\mathbf{A}$ has an eigenvalue λ and an associated eigenvector $\mathbf{x}$ if for $\mathbf{x} \neq \mathbf{0}$, $\mathbf{Ax} = \lambda\mathbf{x}$. It can be shown that if $\mathbf{A}$ is positive-definite and symmetric and $\mathbf{x}$ and $\mathbf{y}$ are distinct eigenvectors, then they are mutually orthogonal, i.e.

$$(\mathbf{x}, \mathbf{y}) = 0 = (\mathbf{y}, \mathbf{x}).$$

Since $(\mathbf{y}, \mathbf{Ax}) = (\mathbf{y}, \lambda\mathbf{x}) = \lambda(\mathbf{y}, \mathbf{x}) = 0$, it follows that the eigenvectors of a positive-definite matrix $\mathbf{A}$ are mutually conjugate with respect to $\mathbf{A}$. Hence, given any positive-definite matrix $\mathbf{A}$ then there exists at least one pair of mutually conjugate directions with respect to this matrix. It is now shown that a set of mutually conjugate vectors in $\mathbb{R}^n$ forms a basis and thus spans $\mathbb{R}^n$.

Theorem 6.5.1

Let $\mathbf{u}^i$, $i = 1, 2, \ldots, n$ be a set of vectors in $\mathbb{R}^n$ which are mutually conjugate with respect to a given symmetric positive-definite matrix $\mathbf{A}$. Then for each $\mathbf{x} \in \mathbb{R}^n$ it holds that

$$\mathbf{x} = \sum_{i=1}^{n} \lambda_i \mathbf{u}^i \text{ where } \lambda_i = \frac{(\mathbf{u}^i, \mathbf{Ax})}{(\mathbf{u}^i, \mathbf{Au}^i)}.$$

Proof

Consider the linear combination $\sum_{i=1}^n \alpha_i \mathbf{u}^i = \mathbf{0}$. Then

$$\mathbf{A}\left(\sum_{i=1}^n \alpha_i \mathbf{u}^i\right) = \sum_{i=1}^n \alpha_i \mathbf{Au}^i = \mathbf{0}.$$

Since the vectors $\mathbf{u}^i$ are mutually conjugate with respect to $\mathbf{A}$ it follows that

$$(\mathbf{u}^k, \mathbf{A}(\textstyle\sum_{i=1}^n \alpha_i \mathbf{u}^i)) = \alpha_k(\mathbf{u}^k, \mathbf{Au}^k) = 0.$$

Since $\mathbf{A}$ is positive-definite and $\mathbf{u}^k \neq \mathbf{0}$ it follows that $\left(\mathbf{u}^k, \mathbf{Au}^k\right) \neq 0$, and thus $\alpha_k = 0$, $k = 1, 2, \ldots, n$. The set $\mathbf{u}^i$, $i = 1, 2, \ldots, n$ thus forms a linear independent set of vectors in $\mathbb{R}^n$ which may be used as a basis.

Thus for any $\mathbf{x}$ in $\mathbb{R}^n$ there exists a unique set λ_i, $i = 1, 2, \ldots, n$ such that

$$\mathbf{x} = \sum_{i=1}^{n} \lambda_i \mathbf{u}^i. \tag{6.23}$$

Now since the $\mathbf{u}^i$ are mutually conjugate with respect to $\mathbf{A}$ it follows that $(\mathbf{u}^i, \mathbf{A}\mathbf{x}) = (\lambda_i \mathbf{u}^i, \mathbf{A}\mathbf{u}^i)$ giving

$$\lambda_i = \frac{(\mathbf{u}^i, \mathbf{A}\mathbf{x})}{(\mathbf{u}^i, \mathbf{A}\mathbf{u}^i)} \tag{6.24}$$

which completes the proof. □

The following lemma is required in order to show that the Fletcher-Reeves directions given by $\mathbf{u}^1 = -\mathbf{g}^0$ and formulae (2.12) and (2.13) are mutually conjugate. Here the notation $\mathbf{g}^k \equiv \boldsymbol{\nabla} f(\mathbf{x}^k)$ is also used.

Lemma 6.5.2

Let $\mathbf{u}^1$, $\mathbf{u}^2$, …, $\mathbf{u}^n$ be mutually conjugate directions with respect to A along an optimal descent path applied to $f(\mathbf{x})$ given by (2.7). Then

$$(\mathbf{u}^k, \mathbf{g}^i) = 0, \quad k = 1, 2, \ldots, i; \quad 1 \leq i \leq n.$$

Proof

For optimal decrease at step k it is required that

$$(\mathbf{u}^k, \boldsymbol{\nabla} f(\mathbf{x}^k)) = 0, \quad k = 1, 2, \ldots, i.$$

Also

$$\begin{aligned}
\boldsymbol{\nabla} f(\mathbf{x}^i) &= \mathbf{A}\mathbf{x}^i + \mathbf{b} \\
&= \left(\mathbf{A}\mathbf{x}^k + \mathbf{A} \sum_{j=k+1}^{i} \lambda_j \mathbf{u}^j \right) + \mathbf{b} \\
&= \boldsymbol{\nabla} f(\mathbf{x}^k) + \sum_{j=k+1}^{i} \lambda_j (\mathbf{A}\mathbf{u}^j)
\end{aligned}$$

and thus

$$(\mathbf{u}^k, \boldsymbol{\nabla} f(\mathbf{x}^i)) = (\mathbf{u}^k, \boldsymbol{\nabla} f(\mathbf{x}^k)) + \sum_{j=k+1}^{i} \lambda_j (\mathbf{u}^k, \mathbf{A}\mathbf{u}^j) = 0$$

which completes the proof. □

Theorem 6.5.3

The directions $\mathbf{u}^i$, $i = 1, 2, \ldots, n$ of the Fletcher-Reeves algorithm given in Section 2.3.2.4 are mutually conjugate with respect to $\mathbf{A}$ of $f(\mathbf{x})$ given by (2.7).

Proof

The proof is by induction.

First, $\mathbf{u}^1$ and $\mathbf{u}^2$ are mutually conjugate:

$$\begin{aligned}(\mathbf{u}^2, \mathbf{A}\mathbf{u}^1) &= -\left((\mathbf{g}^1 + \beta_1\mathbf{g}^0), \mathbf{A}(\mathbf{x}^1 - \mathbf{x}^0)\tfrac{1}{\lambda_1}\right)\\ &= -\left((\mathbf{g}^1 + \beta_1\mathbf{g}^0), (\mathbf{g}^1 - \mathbf{g}^0)\tfrac{1}{\lambda_1}\right)\\ &= -\tfrac{1}{\lambda_1}\left(\|\mathbf{g}^1\|^2 - \beta_1\|\mathbf{g}^0\|^2\right)\\ &= 0.\end{aligned}$$

Now assume that $\mathbf{u}^1$, $\mathbf{u}^2$, ..., $\mathbf{u}^i$ are mutually conjugate, i.e.

$$(\mathbf{u}^k, \mathbf{A}\mathbf{u}^j) = 0, \quad k \neq j, \; k, j \leq i.$$

It is now required to prove that $(\mathbf{u}^k, \mathbf{A}\mathbf{u}^{i+1}) = 0$ for $k = 1, 2, \ldots, i$.

First consider $-(\mathbf{g}^k, \mathbf{g}^i)$ for $k = 1, 2, \ldots, i-1$:

$$\begin{aligned}-(\mathbf{g}^k, \mathbf{g}^i) &= (-\mathbf{g}^k + \beta_k\mathbf{u}^k, \mathbf{g}^i) \quad \text{from Lemma 6.5.2}\\ &= (\mathbf{u}^{k+1}, \mathbf{g}^i) = 0 \quad \text{also from Lemma 6.5.2.}\end{aligned}$$

Hence

$$(\mathbf{g}^k, \mathbf{g}^i) = 0, \quad k = 1, 2, \ldots, i-1. \tag{6.25}$$

Now consider $(\mathbf{u}^k, \mathbf{A}\mathbf{u}^{i+1})$ for $k = 1, 2, \ldots, i-1$:

$$\begin{aligned}(\mathbf{u}^k, \mathbf{A}\mathbf{u}^{i+1}) &= -(\mathbf{u}^k, \mathbf{A}\mathbf{g}^i - \beta_i\mathbf{A}\mathbf{u}^i)\\ &= -(\mathbf{u}^k, \mathbf{A}\mathbf{g}^i) \quad \text{from the induction assumption}\\ &= -(\mathbf{g}^i, \mathbf{A}\mathbf{u}^k)\\ &= -\left(\mathbf{g}^i, \mathbf{A}\frac{\mathbf{x}^k - \mathbf{x}^{k-1}}{\lambda_k}\right)\\ &= -\tfrac{1}{\lambda_k}(\mathbf{g}^i, \mathbf{g}^k - \mathbf{g}^{k-1}) = 0 \quad \text{from (6.25).}\end{aligned}$$

Hence

$$(\mathbf{u}^k, \mathbf{A}\mathbf{u}^{i+1}) = 0, \quad k = 1, 2, \ldots, i-1. \tag{6.26}$$

All that remains is to prove that $(\mathbf{u}^i, \mathbf{A}\mathbf{u}^{i+1}) = 0$ which implies a β_i such that

$$(\mathbf{u}^i, \mathbf{A}(-\mathbf{g}^i + \beta_i \mathbf{u}^i)) = 0,$$

i.e.

$$-(\mathbf{u}^i, \mathbf{A}\mathbf{g}^i) + \beta_i(\mathbf{u}^i, \mathbf{A}\mathbf{u}^i) = 0$$

or

$$\beta_i = \frac{(\mathbf{g}^i, \mathbf{A}\mathbf{u}^i)}{(\mathbf{u}^i, \mathbf{A}\mathbf{u}^i)}. \tag{6.27}$$

Now

$$\begin{aligned}
(\mathbf{g}^i, \mathbf{A}\mathbf{u}^i) &= \tfrac{1}{\lambda_i}(\mathbf{g}^i, (\mathbf{A}\mathbf{x}^i - \mathbf{x}^{i-1})) \\
&= \tfrac{1}{\lambda_i}(\mathbf{g}^i, \mathbf{g}^i - \mathbf{g}^{i-1}) \\
&= \tfrac{1}{\lambda_i}\|\mathbf{g}^i\|^2
\end{aligned}$$

and

$$\begin{aligned}
(\mathbf{u}^i, \mathbf{A}\mathbf{u}^i) &= \tfrac{1}{\lambda_i}(\mathbf{u}^i, \mathbf{A}(\mathbf{x}^i - \mathbf{x}^{i-1})) \\
&= \tfrac{1}{\lambda_i}(\mathbf{u}^i, \mathbf{g}^i - \mathbf{g}^{i-1}) \\
&= -\tfrac{1}{\lambda_i}(\mathbf{u}^i, \mathbf{g}^{i-1}) \quad \text{from Lemma 6.5.2} \\
&= -\tfrac{1}{\lambda_i}(-\mathbf{g}^{i-1} + \beta_{i-1}\mathbf{u}^{i-1}, \mathbf{g}^{i-1}) \\
&= \tfrac{1}{\lambda_i}\|\mathbf{g}^{i-1}\|^2.
\end{aligned}$$

Thus from (6.27)it is required that $\beta_i = \dfrac{\|\mathbf{g}^i\|^2}{\|\mathbf{g}^{i-1}\|^2}$ which agrees with the value prescribed by the Fletcher-Reeves algorithm. Consequently

$$(\mathbf{u}^i, \mathbf{A}\mathbf{u}^{i+1}) = 0.$$

Combining this result with (6.26) completes the proof. □

6.6 DFP method

The following two theorems are concerned with the DFP algorithm. The first theorem shows that it is a descent method.

Theorem 6.6.1

If $\mathbf{G}_i$ is positive-definite and $\mathbf{G}_{i+1}$ is calculated using (2.16) then $\mathbf{G}_{i+1}$ is also positive-definite.

Proof

If $\mathbf{G}_i$ is positive-definite and symmetric then there exists a matrix $\mathbf{F}_i$ such that $\mathbf{F}_i^T\mathbf{F}_i = \mathbf{G}_i$, and

$$\begin{aligned}(\mathbf{x}, \mathbf{G}_{i+1}\mathbf{x}) &= (\mathbf{x}, \mathbf{G}_i\mathbf{x}) + \frac{(\mathbf{v}^j, \mathbf{x})^2}{(\mathbf{y}^j, \mathbf{v}^j)} - \frac{(\mathbf{x}, \mathbf{G}_i\mathbf{y}^j)^2}{(\mathbf{y}^j, \mathbf{G}_i\mathbf{y}^j)}, \; j = i+1 \\ &= \frac{(\mathbf{p}^i, \mathbf{p}^i)(\mathbf{q}^i, \mathbf{q}^i) - (\mathbf{p}^i, \mathbf{q}^i)^2}{(\mathbf{q}^i, \mathbf{q}^i)} + \frac{(\mathbf{v}^j, \mathbf{x})^2}{(\mathbf{y}^j, \mathbf{v}^j)}\end{aligned}$$

where $\mathbf{p}^i = \mathbf{F}_i\mathbf{x}$ and $\mathbf{q}^i = \mathbf{F}_i\mathbf{y}^j$.

If $\mathbf{x} \neq \theta\mathbf{y}^j$ for some scalar θ, it follows from the Schwartz inequality that the first term is strictly positive.

For the second term it holds for the denominator that

$$\begin{aligned}(\mathbf{y}^j, \mathbf{v}^j) &= (\mathbf{g}^j - \mathbf{g}^{j-1}, \mathbf{v}^j) \\ &= -(\mathbf{g}^{j-1}, \mathbf{v}^j) \\ &= \lambda_j(\mathbf{g}^{j-1}, \mathbf{G}_{j-1}\mathbf{g}^{j-1}) \\ &= \lambda_{i+1}(\mathbf{g}^i, \mathbf{G}_i\mathbf{g}^i) > 0 \;\; (\lambda_{i+1} > 0 \text{ and } \mathbf{G}_i \text{ positive-definite})\end{aligned}$$

and hence if $\mathbf{x} \neq \theta\mathbf{y}^j$ the second term is non-negative and the right hand side is strictly positive.

Else, if $\mathbf{x} = \theta\mathbf{y}^j$, the first term is zero and we only have to consider the second term:

$$\begin{aligned}\frac{(\mathbf{v}^j, \mathbf{x})^2}{(\mathbf{y}^j, \mathbf{v}^j)} &= \frac{\theta^2(\mathbf{y}^j, \mathbf{v}^j)^2}{(\mathbf{y}^j, \mathbf{v}^j)} \\ &= \theta^2(\mathbf{y}^j, \mathbf{v}^j) \\ &= \lambda_{i+1}\theta^2(\mathbf{g}^i, \mathbf{G}_i\mathbf{g}^i) > 0.\end{aligned}$$

This completes the proof. □

The theorem above is important as it guarantees descent.

For any search direction $\mathbf{u}^{k+1}$:

$$\frac{df}{d\lambda}(\mathbf{x}^k + \lambda\mathbf{u}^{k+1}) = \mathbf{g}^{kT}\mathbf{u}^{k+1}.$$

For the DFP method, $\mathbf{u}^{k+1} = -\mathbf{G}_k\mathbf{g}^k$ and hence at $\mathbf{x}^k(\lambda = 0)$:

$$\frac{df}{d\lambda} = -\mathbf{g}^{kT}\mathbf{G}_k\mathbf{g}^k.$$

Consequently, if $\mathbf{G}_k$ is positive-definite, descent is guaranteed at $\mathbf{x}^k$, for $\mathbf{g}^k \neq \mathbf{0}$.

Theorem 6.6.2

If the DFP method is applied to a quadratic function of the form given by (2.7), then the following holds:

(i) $(\mathbf{v}^i, \mathbf{A}\mathbf{v}^j) = 0$, $1 \leq i < j \leq k$, $k = 2, 3, \ldots, n$

(ii) $\mathbf{G}_k\mathbf{A}\mathbf{v}^i = \mathbf{v}^i$, $1 \leq i \leq k$, $k = 1, 2, \ldots, n$

where the vectors that occur in the DFP algorithm (see Section 2.4.2.1) are defined by:

$$\begin{aligned}
\mathbf{x}^{i+1} &= \mathbf{x}^i + \lambda_{i+1}\mathbf{u}^{i+1}, \quad i = 0, 1, 2, \ldots \\
\mathbf{u}^{i+1} &= -\mathbf{G}_i\mathbf{g}^i \\
\mathbf{v}^{i+1} &= \lambda_{i+1}\mathbf{u}^{i+1} \\
\mathbf{y}^{i+1} &= \mathbf{g}^{i+1} - \mathbf{g}^i.
\end{aligned}$$

Proof

The proof is by induction. The most important part of the proof, the induction step, is presented. (Prove the initial step yourself. Property (ii) will hold for $k = 1$ if $\mathbf{G}_1\mathbf{A}\mathbf{v}^1 = \mathbf{v}^1$. Show this by direct substitution. The first case $(\mathbf{v}^1, \mathbf{A}\mathbf{v}^2) = 0$ in (i) corresponds to $k = 2$ and follows from the fact that (ii) holds for $k = 1$. That (ii) also holds for $k = 2$, follows from the second part of the induction proof given below.)

Assume that (i) and (ii) hold for k. Then it is required to show that they also hold for $k + 1$.

First part: proof that (i) is true for $k+1$. For the quadratic function:

$$\mathbf{g}^k = \mathbf{b} + \mathbf{A}\mathbf{x}^k = \mathbf{b} + \mathbf{A}\mathbf{x}^i + \mathbf{A}\sum_{j=i+1}^{k} \mathbf{v}^j.$$

Now consider:

$$(\mathbf{v}^i, \mathbf{g}^k) = (\mathbf{v}^i, \mathbf{g}^i) + \sum_{j=i+1}^{k} (\mathbf{v}^i, \mathbf{A}\mathbf{v}^j) = 0, \quad 1 \le i \le k.$$

The first term on the right is zero from the optimal descent property and the second term as a consequence of the induction assumption that (i) holds for k.

Consequently, with $\mathbf{v}^{k+1} = \lambda_{k+1}\mathbf{u}^{k+1} = -\lambda_{k+1}\mathbf{G}_k\mathbf{g}^k$ it follows that

$$(\mathbf{v}^i, \mathbf{A}\mathbf{v}^{k+1}) = -\lambda_{k+1}(\mathbf{v}^i, \mathbf{A}\mathbf{G}_k\mathbf{g}^k) = -\lambda_{k+1}(\mathbf{G}_k\mathbf{A}\mathbf{v}^i, \mathbf{g}^k), \quad \lambda_{k+1} > 0.$$

Hence

$$(\mathbf{v}^i, \mathbf{A}\mathbf{v}^{k+1}) = -\lambda_{k+1}(\mathbf{v}^i, \mathbf{g}^k) = 0$$

as a consequence of the induction assumption (ii) and the result above.

Hence with $(\mathbf{v}^i, \mathbf{A}\mathbf{v}^{k+1}) = 0$, property (i) holds for $k+1$.

Second part: proof that (ii) holds for $k+1$. Furthermore

$$\begin{aligned}
(\mathbf{y}^{k+1}, \mathbf{G}_k\mathbf{A}\mathbf{v}^i) &= (\mathbf{y}^{k+1}, \mathbf{v}^i) \quad \text{for } i \le k \text{ from assumption (ii)} \\
&= (\mathbf{g}^{k+1} - \mathbf{g}^k, \mathbf{v}^i) \\
&= (\mathbf{A}(\mathbf{x}^{k+1} - \mathbf{x}^k), \mathbf{v}^i) \\
&= (\mathbf{A}\mathbf{v}^{k+1}, \mathbf{v}^i) \\
&= (\mathbf{v}^i, \mathbf{A}\mathbf{v}^{k+1}) = 0 \text{ from the first part.}
\end{aligned}$$

Using the update formula, it follows that

$$\begin{aligned}
\mathbf{G}_{k+1}\mathbf{A}\mathbf{v}^i &= \mathbf{G}_k\mathbf{A}\mathbf{v}^i + \frac{\mathbf{v}^{k+1}(\mathbf{v}^{k+1})^T\mathbf{A}\mathbf{v}^i}{(\mathbf{v}^{k+1})^T\mathbf{v}^{k+1}} - \frac{(\mathbf{G}_k\mathbf{y}^{k+1})(\mathbf{G}_k\mathbf{y}^{k+1})^T\mathbf{A}\mathbf{v}^i}{(\mathbf{y}^{k+1})^T\mathbf{G}_k\mathbf{y}^{k+1}} \\
&= \mathbf{G}_k\mathbf{A}\mathbf{v}^i \;\; (\text{because } (\mathbf{v}^i, \mathbf{A}\mathbf{v}^{k+1}) = 0,\ (\mathbf{y}^{k+1}, \mathbf{G}_k\mathbf{A}\mathbf{v}^i) = 0) \\
&= \mathbf{v}^i \quad \text{for } i \le k \quad \text{from assumption (ii).}
\end{aligned}$$

It is still required to show that $\mathbf{G}_{k+1}\mathbf{A}\mathbf{v}^{k+1} = \mathbf{v}^{k+1}$. This can be done, as for the initial step where it was shown that $\mathbf{G}_1\mathbf{A}\mathbf{v}^1 = \mathbf{v}^1$, by direct substitution. □

Thus it was shown that $\mathbf{v}^k$, $k = 1, \ldots, n$ are mutually conjugate with respect to $\mathbf{A}$ and therefore they are linearly independent and form a basis for $\mathbb{R}^n$. Consequently the DFP method is quadratically terminating with $\mathbf{g}^n = \mathbf{0}$. Property (ii) also implies that $\mathbf{G}_n = \mathbf{A}^{-1}$.

A final interesting result is the following theorem.

Theorem 6.6.3

If the DFP method is applied to $f(\mathbf{x})$ given by (2.7), then

$$\mathbf{A}^{-1} = \sum_{i=1}^{n} \mathbf{A}_i.$$

Proof

In the previous proof it was shown that the $\mathbf{v}^i$ vectors , $i = 1, 2, \ldots, n$, are mutually conjugate with respect to $\mathbf{A}$. They therefore form a basis in $\mathbb{R}^n$. Also

$$\mathbf{A}_i = \frac{\mathbf{v}^i\mathbf{v}^{iT}}{\mathbf{v}^{iT}\mathbf{y}^i} = \frac{\mathbf{v}^i\mathbf{v}^{iT}}{\mathbf{v}^{iT}(\mathbf{g}^i - \mathbf{g}^{i-1})} = \frac{\mathbf{v}^i\mathbf{v}^{iT}}{\mathbf{v}^{iT}\mathbf{A}\mathbf{v}^i}.$$

Let

$$\mathbf{B} = \sum_{i=1}^{n} \mathbf{A}_i = \sum_{i=1}^{n} \frac{\mathbf{v}^i\mathbf{v}^{iT}}{\mathbf{v}^{iT}\mathbf{A}\mathbf{v}^i}.$$

Then

$$\mathbf{B}\mathbf{A}\mathbf{x} = \sum_{i=1}^{n} \frac{\mathbf{v}^i\mathbf{v}^{iT}}{\mathbf{v}^{iT}\mathbf{A}\mathbf{v}^i}\mathbf{A}\mathbf{x} = \sum_{i=1}^{n} \frac{\mathbf{v}^{iT}\mathbf{A}\mathbf{x}}{\mathbf{v}^{iT}\mathbf{A}\mathbf{v}^i}\mathbf{v}^i = \mathbf{x}$$

as a consequence of Theorem 6.5.1.

This result holds for arbitrary $\mathbf{x} \in \mathbb{R}^n$ and hence $\mathbf{B}\mathbf{A} = \mathbf{I}$ where $\mathbf{I}$ is the identity matrix, and it follows that

$$\mathbf{B} = \sum_{i=1}^{n} \mathbf{A}_i = \mathbf{A}^{-1}.$$

□

Appendix A

THE SIMPLEX METHOD FOR LINEAR PROGRAMMING PROBLEMS

A.1 Introduction

This introduction to the simplex method is along the lines given by Chvatel (1983).

Here consider the *maximization* problem:

$$\begin{aligned} &\text{maximize } Z = \mathbf{c}^T\mathbf{x} \\ &\text{such that } \mathbf{Ax} \leq \mathbf{b},\ \mathbf{A} \text{ an } m \times n \text{ matrix} \\ &x_i \geq 0,\ i = 1, 2, ..., n. \end{aligned} \tag{A.1}$$

Note that $\mathbf{Ax} \leq \mathbf{b}$ is equivalent to $\sum_{i=1}^{n} a_{ji}x_i \leq b_j,\ j = 1, 2, ..., m.$

Introduce *slack variables* $x_{n+1}, x_{n+2}, ..., x_{n+m} \geq 0$ to transform the inequality constraints to equality constraints:

$$\begin{aligned} a_{11}x_1 + \ldots \quad + \quad & a_{1n}x_n + x_{n+1} = b_1 \\ a_{21}x_1 + \ldots \quad + \quad & a_{2n}x_n + x_{n+2} = b_2 \\ & \vdots \\ a_{m1}x_1 + \ldots \quad + \quad & a_{mn}x_n + x_{n+m} = b_m \end{aligned} \tag{A.2}$$

or

$$[\mathbf{A}; \mathbf{I}]\mathbf{x} = \mathbf{b}$$

where $\mathbf{x} = [x_1, x_2, ..., x_{n+m}]^T, \mathbf{b} = [b_1, b_2, ..., b_m]^T$, and $x_1, x_2, ..., x_n \geq 0$ are the *original decision variables* and $x_{n+1}, x_{n+2}, ..., x_{n+m} \geq 0$ the *slack variables.*

Now assume that $b_i \geq 0$ for all i, To start the process an initial *feasible* solution is then given by:

$$\begin{aligned} x_{n+1} &= b_1 \\ x_{n+2} &= b_2 \\ &\vdots \\ x_{n+m} &= b_m \end{aligned}$$

with $x_1 = x_2 = \cdots = x_n = 0$.

In this case we have a *feasible origin.*

We now write system (A.2) in the so called standard *tableau* format:

$$\begin{aligned} x_{n+1} &= b_1 - a_{11}x_1 - \ldots - a_{1n}x_n \geq 0 \\ x_{n+2} &= b_2 - a_{21}x_1 - \ldots - a_{2n}x_n \geq 0 \\ &\vdots \\ x_{n+m} &= b_m - a_{m1}x_1 - \ldots - a_{mn}x_n \geq 0 \\ Z &= c_1x_1 + c_2x_2 + \ldots + c_nx_n \end{aligned} \tag{A.3}$$

The left side contains the *basic variables*, in general $\neq 0$, and the right side the *nonbasic variables*, all $= 0$. The last line Z denotes the objective function (in terms of *nonbasic* variables).

In a more general form the *tableau* can be written as

$$\begin{aligned} x_{B1} &= b_1 - a_{11}x_{N1} - \ldots - a_{1n}x_{Nn} \geq 0 \\ x_{B2} &= b_2 - a_{21}x_{N1} - \ldots - a_{2n}x_{Nn} \geq 0 \\ &\vdots \\ x_{Bm} &= b_m - a_{m1}x_{N1} - \ldots - a_{mn}x_{Nn} \geq 0 \\ Z &= c_{N1}x_{N1} + c_{N2}x_{N2} + \ldots + c_{Nn}x_{Nn} \end{aligned} \tag{A.4}$$

The $\geq$ at the right serves to remind us that $x_{Bj} \geq 0$ is a necessary condition, even when the values of x_{Ni} change from their zero values.

$\mathbf{x}_B$ = vector of basic variables and $\mathbf{x}_N$ = vector of nonbasic variables represent a *basic feasible solution.*

A.2 Pivoting for increase in objective function

Clearly if any $c_{Np} > 0$, then Z increases if we increase x_{Np}, with the other $x_{Ni} = 0, i \neq p$. Assume further that $c_{Np} > 0$ and $c_{Np} \geq c_{Ni}, i = 1, 2, ..., n$, then we decide to increase x_{Np}. But x_{Np} can not be increased indefinitely because of the constraint $\mathbf{x}_B \geq \mathbf{0}$ in *tableau* (A.4). Every entry i in the tableau, with $a_{ip} > 0$, yields a constraint on x_{Np} of the form:

$$0 \leq x_{Np} \leq \frac{b_i}{a_{ip}} = d_i, \quad i = 1, 2, ..., m. \tag{A.5}$$

Assume now that $i = k$ yields the strictest constraint, then let $x_{Bk} = 0$ and $x_{Np} = d_k$. Now $x_{Bk}(= 0)$ is the *outgoing* variable (out of the base), and $x_{Np} = d_k(\neq 0)$ the *incoming* variable. The k-th entry in tableau (A.4) changes to

$$x_{Np} = d_k - \sum_{i \neq k} \bar{a}_{ki}x_{Ni} - \bar{a}_{kk}x_{Bk} \tag{A.6}$$

with $\bar{a}_{ki} = a_{ki}/a_{kp}$, $\bar{a}_{kk} = 1/a_{kp}$.

Replace x_{Np} by (A.6) in each of the remaining $m - 1$ entries in tableau (A.4) as well as in the objective function Z. With (A.6) as the first entry this gives the *new tableau* in terms of the new basic variables:

$$x_{B1}, x_{B2}, \ldots, x_{Np}, ..., x_{Bm} \text{ (left side) } \neq 0$$

and nonbasic variables:

$$x_{N1}, x_{N2}, \ldots, x_{Bk}, \ldots, x_{Nn} \text{ (right side)} = 0.$$

As x_{Np} has increased with d_k, the objective function has also increased by $c_{Np}d_k$. The objective function (last) entry is thus of the form

$$Z = c_{Np}d_k + c_{N1}x_{N1} + c_{N2}x_{N2} + \cdots + c_{Nm}x_{Nm}$$

where the x_{Ni} now denotes the new nonbasic variables and c_{Ni} the new associated coefficients. This completes the first *pivoting iteration.*

Repeat the procedure above until a tableau is obtain such that

$$Z = Z^* + c_{N1}x_{N1} + \cdots + c_{Nm}x_{Nm} \text{ with } c_{Ni} \leq 0, \quad i = 1, 2, ..., m.$$

The optimal value of the objective function is then $Z = Z^*$ (no further increase is possible).

A.3 Example

maximize $Z = 5x_1 + 4x_2 + 3x_3$ such that

$2x_1 + 3x_2 + x_3 \leq 5$
$4x_1 + x_2 + 2x_3 \leq 11$
$3x_1 + 4x_2 + 2x_3 \leq 8$
$x_1, x_2, x_3 \geq 0.$

This problem has a *feasible origin.* Introduce slack variables x_4, x_5 and x_6 and then the *first tableau* is given by:

I:

$$\begin{array}{ccccccccc|l} x_4 & = & 5 & - & 2x_1 & - & 3x_2 & - & x_3 \geq 0 & x_1 \leq 5/2(s) \\ x_5 & = & 11 & - & 4x_1 & - & x_2 & - & 2x_3 \geq 0 & x_1 \leq 11/4 \\ x_6 & = & 8 & - & 3x_1 & - & 4x_2 & - & 2x_3 \geq 0 & x_1 \leq 8/3 \\ \hline Z & = & & & 5x_1 & + & 4x_2 & + & 3x_3 & \end{array}$$

Here $5 > 0$ and $5 > 4 > 3$. Choose thus x_1 as *incoming variable.* To find the outgoing variable, calculate the constraints on x_1 for all the entries (see right side). The strictest (s) constraint is given by the first entry, and thus the *outgoing variable* is x_4. The first entry in the next tableau is

$$x_1 = \tfrac{5}{2} - \tfrac{3}{2}x_2 - \tfrac{1}{2}x_3 - \tfrac{1}{2}x_4.$$

Replace this expression for x_1 in all other entries to find the next tableau:

$$\begin{array}{rcll}
x_1 & = & \frac{5}{2} - \frac{3}{2}x_2 - \frac{1}{2}x_3 - \frac{1}{2}x_4 & \geq 0 \\
x_5 & = & 11 - 4(\frac{5}{2} - \frac{3}{2}x_2 - \frac{1}{2}x_3 - \frac{1}{2}x_4) - x_2 - 2x_3 & \geq 0 \\
x_6 & = & 8 - 3(\frac{5}{2} - \frac{3}{2}x_2 - \frac{1}{2}x_3 - \frac{1}{2}x_4) - x4x_2 - 2x_3 & \geq 0 \\
\hline
Z & = & 5(\frac{5}{2} - \frac{3}{2}x_2 - \frac{1}{2}x_3 - \frac{1}{2}x_4) + 4x_2 + 3x_3 &
\end{array}$$

After simplification we obtain the *second tableau* in standard format:

II:

$$\begin{array}{rcccccccccl|l}
x_1 & = & \frac{5}{2} & - & \frac{3}{2}x_2 & - & \frac{1}{2}x_3 & - & \frac{1}{2}x_4 & \geq 0 & x_3 \leq 5 \\
x_5 & = & 1 & + & 5x_2 & & & + & 2x_4 & \geq 0 & \text{no bound} \\
x_6 & = & \frac{1}{2} & + & \frac{1}{2}x_2 & - & \frac{1}{2}x_3 & + & \frac{3}{2}x_4 & \geq 0 & x_3 \leq 1(s) \\
\hline
Z & = & \frac{25}{5} & - & \frac{7}{2}x_2 & + & \frac{1}{2}x_3 & - & \frac{5}{2}x_4 & &
\end{array}$$

This completes the first iteration. For the next step it is clear that x_3 is the incoming variable and consequently the outgoing variable is x_6. The first entry for the next tableau is thus $x_3 = 1 + x_2 + 3x_4 - 2x_6$ (3-rd entry in previous tableau).

Replace this expression for x_3 in all the remaining entries of tableau II. After simplification we obtain the *third tableau*:

III:

$$\begin{array}{rcccccccccc}
x_3 & = & 1 & + & x_2 & + & 3x_4 & - & 2x_6 & \geq & 0 \\
x_1 & = & 2 & - & 2x_2 & - & 2x_4 & + & x_6 & \geq & 0 \\
x_5 & = & 1 & + & 5x_2 & + & 2x_4 & & & \geq & 0 \\
\hline
Z & = & 13 & - & 3x_2 & - & x_4 & - & x_6 & &
\end{array}$$

In the last entry all the coefficients of the nonbasic variables are negative. Consequently it is not possible to obtain a further increase in Z by increasing one of the nonbasic variables. The optimal value of Z is thus $Z^* = 13$ with

$$x_1^* = 2 \ ; \ \ x_2^* = 0 \ ; \ \ x_3^* = 1.$$

Assignment A.1

Solve by using the simplex method:

maximize $z = 3x_1 + 2x_2 + 4x_3$ such that

$x_1 + x_2 + 2x_3 \leq 4$
$2x_1 + 3x_3 \leq 5$
$2x_1 + x_2 + 3x_3 \leq 7$
$x_1, x_2, x_3 \geq 0.$

A.4 The auxiliary problem for problem with infeasible origin

In the previous example it is possible to find the solution using the simplex method only because $b_i > 0$ for all i and an initial solution $x_i = 0 \ , \ i = 1, 2, ...n$ with $x_{n+j} = b_j, \ \ j = 1, 2, ..., m$ was thus feasible, that is, the origin is a feasible initial solution.

If the LP problem does not have a feasible origin we first solve the so called *auxiliary problem*:

Phase 1:

$$\begin{aligned} &\text{maximize } W = -x_0 \\ \text{such that } &\textstyle\sum_{i=1}^{n} a_{ji}x_i - x_0 \leq b_j, \quad j = 1, 2, ..., m \\ &x_i \geq 0, \quad i = 0, 1, 2, ...n \end{aligned} \tag{A.7}$$

where x_0 is called the new artificial variable. By setting $x_i = 0$ for $i = 1, 2, ...n$ and choosing x_0 large enough, we can always find a feasible solution.

The original problem clearly has a feasible solution if and only if the auxiliary problem has a feasible solution with $x_0 = 0$ or, in other words, the original problem has a feasible solution if and only if the optimal value of the auxiliary problem is zero. The original problem is now solved using the simplex method, as described in the previous sections. This solution is called *Phase* 2.

A.5 Example of auxiliary problem solution

Consider the LP:

maximize $Z = x_1 - x_2 + x_3$ such that

$2x_1 - x_2 + x_3 \leq 4$
$2x_1 - 3x_2 + x_3 \leq -5$
$-x_1 + x_2 - 2x_3 \leq -1$
$x_1, x_2, x_3 \geq 0.$

Clearly this problem does not have a feasible origin.

We first perform *Phase* 1:

Consider the *auxiliary problem:*

maximize $W = -x_0$ such that

$2x_1 - x_2 + 2x_3 - x_0 \leq 4$
$2x_1 - 3x_2 + x_3 - x_0 \leq -5$
$-x_1 + x_2 - 2x_3 - x_0 \leq -1$
$x_0, x_1, x_2, x_3 \geq 0.$

Introduce the slack variables x_4, x_5 and x_6, which gives the tableau (not yet in standard form):

x_4	$=$	4	$-$	$2x_1$	$+$	x_2	$-$	$2x_3$	$+$	x_0	$\geq$	0	$x_0 \geq -4$
x_5	$=$	-5	$-$	$2x_1$	$+$	$3x_2$	$-$	x_3	$+$	x_0	$\geq$	0	$x_0 \geq 5(s)$
x_6	$=$	-1	$+$	x_1	$-$	x_2	$+$	$2x_3$	$+$	x_0	$\geq$	0	$x_0 \geq 1$
W	$=$								$-$	x_0			

This is not in standard form as x_0 on the right is not zero. With $x_1 = x_2 = x_3 = 0$ then $x_4, x_5, x_6 \geq 0$ if $x_0 \geq \max\{-4; 5; 1\}$.

Choose $x_0 = 5$, as prescribed by the second (strictest) entry. This gives $x_5 = 0, x_4 = 9$ and $x_6 = 4$. Thus x_0 is a basic variable ($\neq 0$) and x_5 a nonbasic variable. The *first standard tableau* can now be write as

I:

$$
\begin{array}{rcrrrrrr|l}
x_0 & = & 5 & +\ 2x_1 & -\ 3x_2 & +\ x_3 & +\ x_5 & \geq 0 & x_2 \leq \frac{5}{3} \\
x_4 & = & 9 & & -\ 2x_2 & -\ x_3 & +\ x_5 & \geq 0 & x_2 \leq \frac{9}{2} \\
x_6 & = & 4 & +\ 3x_1 & -\ 4x_2 & +\ 3x_3 & +\ x_5 & \geq 0 & x_2 \leq 1(s) \\
\hline
W & = & -5 & -\ 2x_1 & +\ 3x_2 & -\ x_3 & -\ x_5 & &
\end{array}
$$

Now apply the simplex method. From the last entry it is clear that W increases as x_2 increases. Thus x_2 is the *incoming variable.* With the strictest bound $x_2 \leq 1$ as prescribed by the third entry the *outgoing variable* is x_6. The second tableau is given by:

II:

$$
\begin{array}{rcrrrrr|l}
x_2 & = & 1+ & 0.75x_1+ & 0.75x_3+ & 0.25x_5- & 0.25x_6 \geq 0 & \text{no bound} \\
x_0 & = & 2- & 0.25x_1- & 1.25x_3+ & 0.25x_5- & 0.75x_6 \geq 0 & x_3 \leq \frac{8}{5}(s) \\
x_4 & = & 7- & 1.5x_1- & 2.5x_3+ & 0.5x_5+ & 0.5x_6 \geq 0 & x_3 \leq \frac{14}{5} \\
\hline
W & = & -2+ & 0.25x_1+ & 1.25x_3- & 0.25x_5- & 0.75x_6 &
\end{array}
$$

The new incoming variable is x_3 and the outgoing variable x_0. Perform the necessary pivoting and simplify. The next tableau is then given by:

III:

$$
\begin{array}{rcrrrrrr}
x_3 & = & 1.6 & -\ 0.2x_1 & +\ 0.2x_5 & +\ 0.6x_6 & -\ 0.8x_0 & \geq 0 \\
x_2 & = & 2.2 & +\ 0.6x_1 & +\ 0.4x_5 & +\ 0.2x_6 & -\ 0.6x_0 & \geq 0 \\
x_4 & = & 3 & -\ x_1 & & -\ x_6 & +\ 2x_0 & \geq 0 \\
\hline
W & = & & & & & -\ x_0 &
\end{array}
$$

As the coefficients of x_0 in the last entry is negative, no further increase in W is possible. Also, as $x_0 = 0$, the solution

$$x_1 = 0; \quad x_2 = 2,2; \quad x_3 = 1,6; \quad x_4 = 3; \quad x_5 = 0; \quad x_6 = 0$$

corresponds to a feasible solution of the original problem. This means that the first phase has been completed.

The initial tableau for *Phase* 2 is simply the above tableau III without the x_0 terms and with the objective function given by:

$$\begin{aligned} Z &= x_1 - x_2 + x_3 \\ &= x_1 - (2.2 + 0.6x_1 + 0.4x_5 + 0.2x_6) + (1.6 - 0.2x_1 + 0.2x_5 + 0.6x_6) \\ &= -0.6 + 0.2x_1 - 0.2x_5 + 0.4x_6 \end{aligned}$$

in terms of nonbasic variables.

Thus the initial tableau for the original problem is:

$$\begin{array}{rcrcrcrcrcl|l} x_3 & = & 1.6 & - & 0.2x_1 & + & 0.2x_5 & + & 0.6x_6 & \geq & 0 & \text{no bound} \\ x_2 & = & 2.2 & + & 0.6x_1 & + & 0.4x_5 & + & 0.2x_6 & \geq & 0 & \text{no bound} \\ x_4 & = & 3 & - & x_1 & & & - & x_6 & \geq & 0 & x_6 \leq 3(s) \\ \hline Z & = & -0.6 & + & 0.2x_1 & - & 0.2x_5 & + & 0.4x_6 & & & \end{array}$$

Perform the remaining iterations to find the final solution (the next incoming variable is x_6 with outgoing variable x_4).

Assignment A.2

Solve the following problem using the *two phase* simplex method:

maximize $Z = 3x_1 + x_2$ such that

$$\begin{aligned} &x_1 - x_2 \leq -1 \\ &-x_1 - x_2 \leq -3 \\ &2x_1 + x_2 \leq 4 \\ &x_1, x_2 \geq 0. \end{aligned}$$

A.6 Degeneracy

A further complication that may occur is degeneracy. It is possible that there is more than one candidate outgoing variable. Consider, for example, the following tableau:

$$\begin{array}{rcrcrcrcrcc|l}
x_4 & = & 1 & & & & & - & 2x_3 & \geq & 0 & x_3 \leq \frac{1}{2}(s) \\
x_5 & = & 3 & - & 2x_1 & + & 4x_2 & - & 6x_3 & \geq & 0 & x_3 \leq \frac{1}{2}(s) \\
x_6 & = & 2 & + & x_1 & - & 3x_2 & - & 4x_3 & \geq & 0 & x_3 \leq \frac{1}{2}(s) \\
\hline
Z & = & & & 2x_1 & - & x_2 & + & 8x_3 & & &
\end{array}$$

With x_3 the incoming variable there are three candidates, x_4, x_5 and x_6, for outgoing variable. Choose arbitrarily x_4 as the outgoing variable. Then the tableau is:

$$\begin{array}{rcrcrcrcrcc|l}
x_3 & = & 0.5 & & & & & - & 0.5x_4 & \geq & 0 & \text{no bound on } x_1 \\
x_5 & = & & - & 2x_1 & + & 4x_2 & + & 3x_4 & \geq & 0 & x_1 \leq 0(s) \\
x_6 & = & & & x_1 & - & 3x_2 & + & 2x_4 & \geq & 0 & x_1 \leq 0(s) \\
\hline
Z & = & 4 & + & 2x_1 & - & x_2 & - & 4x_4 & & &
\end{array}$$

This tableau differs from the previous tableaus in one important way: two *basic* variables have the value *zero*. A basic feasible solution for which one or more or the basic variables are zero, is called a *degenerate* solution. This may have bothersome consequences. For example, for the next iteration in our example, with x_1 as the incoming variable and x_5 the outgoing variable there is no increase in the objective function. Such an iteration is called a degenerate iteration. Test the further application of the simplex method to the example for yourself. Usually the stalemate is resolved after a few degenerate iterations and the method proceeds to the optimal solution.

In some, very exotic, cases it may happen that the stalemate is not resolved and the method gets stuck in an infinite loop without any progress towards a solution. So called cycling then occurs. More information on this phenomenon can be obtained in the book by Chvatel (1983).

A.7 The revised simplex method

The revised simplex method (RSM) is equivalent to the ordinary simplex method in terms of tableaus except that matrix algebra is used for the calculations and that the method is, in general, faster for large and sparse systems. For these reasons modern computer programs for LP problems always use the RSM.

We introduce the necessary terminology and then give the algorithm for an iteration of the RSM. From an analysis of the RSM it is clear that the algorithm corresponds in essence with the tableau simplex method. The only differences occur in the way in which the calculations are performed to obtain the incoming and outgoing variables, and the new basic feasible solution. In the RSM two linear systems are solved in each iteration. In practice special factorizations are applied to find these solutions in an economic way. Again see Chvatel (1983).

Consider the LP problem:

$$\text{maximize } Z = \mathbf{c}^T\mathbf{x} \tag{A.8}$$

such that $\mathbf{Ax} \leq \mathbf{b}$, $\mathbf{A}$ $m \times n$ and $\mathbf{x} \geq \mathbf{0}$.

After introducing the slack variables the constraints can be written as:

$$\tilde{\mathbf{A}}\mathbf{x} = \mathbf{b} \quad , \quad \mathbf{x} \geq 0 \tag{A.9}$$

where $\mathbf{x}$ includes the slack variables.

Assume that a basic feasible solution is available. Then, if $\mathbf{x}_B$ denotes the m basic variables, and $\mathbf{x}_N$ the n nonbasic variables, (A.9) can be written as:

$$\tilde{\mathbf{A}}\mathbf{x} = [\mathbf{A}_B\mathbf{A}_N]\begin{bmatrix} \mathbf{x}_B \\ \mathbf{x}_N \end{bmatrix} = \mathbf{b}$$

or

$$\mathbf{A}_B\mathbf{x}_B + \mathbf{A}_N\mathbf{x}_N = \mathbf{b} \tag{A.10}$$

where $\mathbf{A}_B$ is an $m \times m$ and $\mathbf{A}_N$ an $m \times n$ matrix.

The objective function Z can be written as

$$Z = \mathbf{c}_B^T\mathbf{x}_B + \mathbf{c}_N^T\mathbf{x}_N \tag{A.11}$$

where $\mathbf{c}_B$ and $\mathbf{c}_N$ are respectively the basic and nonbasic coefficients.

It can be shown that A_B is always non-singular. It thus follows that

$$\mathbf{x}_B = \mathbf{A}_B^{-1}\mathbf{b} - \mathbf{A}_B^{-1}\mathbf{A}_N\mathbf{x}_N. \tag{A.12}$$

Expression (A.12) clearly corresponds, in matrix form, to the first m entries of the ordinary simplex tableau, while the objective function entry is given by

$$\begin{aligned} Z &= \mathbf{c}_B^T\mathbf{x}_B + \mathbf{c}_N^T\mathbf{x}_N \\ &= \mathbf{c}_B^T\left(\mathbf{A}_B^{-1}\mathbf{b} - \mathbf{A}_B^{-1}\mathbf{A}_N\mathbf{x}_N\right) + \mathbf{c}_N^T\mathbf{x}_N \\ &= \mathbf{c}_B^T\mathbf{A}_B^{-1}\mathbf{b} + \left(\mathbf{c}_N^T - \mathbf{c}_B^T\mathbf{A}_B^{-1}\mathbf{A}_N\right)\mathbf{x}_N. \end{aligned}$$

Denote the basis matrix $\mathbf{A}_B$ by $\mathbf{B}$. The complete tableau is then given by

$$\frac{\mathbf{x}_B = \overbrace{\mathbf{B}^{-1}\mathbf{b}}^{\mathbf{x}_B^*} - \mathbf{B}^{-1}\mathbf{A}_N\mathbf{x}_N \geq \mathbf{0}}{Z = \underbrace{\mathbf{c}_B^T\mathbf{B}^{-1}\mathbf{b}}_{Z^*} + \left(\mathbf{c}_N^T - \mathbf{c}_B^T\mathbf{B}^{-1}\mathbf{A}_N\right)\mathbf{x}_N} \tag{A.13}$$

We now give the RSM in terms of the matrix notation introduced above. A careful study of the algorithm will show that this corresponds exactly to the tableau method which we developed by way of introduction.

A.8 An iteration of the RSM

(Chvatal, 1983)

Step 1: Solve the following system:

$\mathbf{y}^T\mathbf{B} = \mathbf{c}_B^T$. This gives $\mathbf{y}^T = \mathbf{c}_B^T\mathbf{B}^{-1}$.

Step 2: Choose an incoming column. This is any column $\mathbf{a}$ of $\mathbf{A}_N$ such that $\mathbf{y}^T\mathbf{a}$ is less than the corresponding component of $\mathbf{c}_N$.

(See (A.13): $Z = Z^* + (\mathbf{c}_N^T - \mathbf{y}^T\mathbf{A}_N)\mathbf{x}_N$)

If no such column exists, the current solution is optimal.

Step 3: Solve the following system:

$\mathbf{B}\mathbf{d} = \mathbf{a}$. This gives $\mathbf{d} = \mathbf{B}^{-1}\mathbf{a}$.

(From (A.13) it follows that $\mathbf{x}_B = \mathbf{x}_B^* - \mathbf{d}t \geq \mathbf{0}$, where t is the value of the incoming variable).

Step 4: Find the largest value of t such that

$$\mathbf{x}_B^* - t\mathbf{d} \geq \mathbf{0}.$$

If no such t exists, then the problem is unbounded; otherwise at least one component of $\mathbf{x}_B^* - t\mathbf{d}$ will equal zero and the corresponding variable is the outgoing variable.

Step 5: Set the incoming variable, (the new basic variable), equal to t and the other remaining basis variables

$$\mathbf{x}_B^* := \mathbf{x}_B^* - t\mathbf{d}$$

and exchange the outgoing column in $\mathbf{B}$ with the incoming column $\mathbf{a}$ in $\mathbf{A}_N$.

Bibliography

J.S. Arora. *Introduction to Optimal Design.* McGraw-Hill, New York, 1989.

J.S. Arora, O.A. El-wakeil, A.I. Chahande, and C.C. Hsieh. Global optimization methods for engineering applications: A review. Preprint, 1995.

J. Barhen, V. Protopopescu, and D. Reister. TRUST: A deterministic algorithm for global optimization. *Science*, 276:1094–1097, May 1997.

M.S. Bazaraa, H.D. Sherali, and C.M. Shetty. *Nonlinear Programming, Theory and Algorithms.* John Wiley, New York, 1993.

D.P. Bertsekas. Multiplier methods: A survey. *Automatica*, 12:133–145, 1976.

F.H. Branin. Widely used convergent methods for finding multiple solutions of simultaneous equations. *IBM J. Research Development.*, pages 504–522, 1972.

F.H. Branin and S.K. Hoo. A method for finding multiple extrema of a function of n variables. In F.A. Lootsma, editor, *Numerical Methods of Nonlinear Optimization*, pages 231–237, London, 1972. Academic Press.

V. Chvatel. *Linear Programming.* W.H. Freeman, New York, 1983.

G. Dantzig. *Linear Programming and Extensions.* Princeton University Press, Princeton, 1963.

W.C. Davidon. Variable metric method for minimization. R and D Report ANL-5990D (Rev.), Atomic Energy Commission, 1959.

E. De Klerk and J. A. Snyman. A feasible descent cone method for linearly constrained minimization problems. *Computers and Mathematics with Applications*, 28:33–44, 1994.

L.C.W. Dixon, J. Gomulka, and S.E. Hersom. Reflections on the global optimization problem. In L.C.W. Dixon, editor, *Optimization in Action*, pages 398–435, London, 1976. Academic Press.

L.C.W. Dixon, J. Gomulka, and G.P. Szegö. Towards a global optimization technique. In L.C.W. Dixon and G.P. Szegö, editors, *Towards Global Optimization*, pages 29–54, Amsterdam, 1975. North-Holland.

L.C.W. Dixon and G.P. Szegö. The global optimization problem: An introduction. In L.C.W. Dixon and G.P. Szegö, editors, *Towards Global Optimization 2*, pages 1–15, Amsterdam, 1978. North-Holland.

J. Farkas and K. Jarmai. *Analysis and Optimum Design of Metal Structures.* A.A. Balkema, Rotterdam, 1997.

A.V. Fiacco and G.P. McCormick. *Nonlinear Programming.* John Wiley, New York, 1968.

R. Fletcher. An ideal penalty function for constrained optimization. *Journal of the Institute of Mathematics and its Applications*, 15:319–342, 1975.

R. Fletcher. *Practical Methods of Optimization.* John Wiley, Chichester, 1987.

R. Fletcher and C.M. Reeves. Function minimization by conjugate gradients. *Computer Journal*, 7:140–154, 1964.

C. Fleury. Structural weight optimization by dual methods of convex programming. *International Journal for Numerical Methods in Engineering*, 14:1761–1783, 1979.

J.C. Gilbert and J. Nocedal. Global convergence properties of conjugate gradient methods. *SIAM Journal on Optimization*, 2:21–42, 1992.

A.O. Griewank. Generalized descent for global optimization. *Journal of Optimization Theory and Applications*, 34:11–39, 1981.

A.A. Groenwold and J.A. Snyman. Global optimization using dynamic search trajectories. *Journal of Global Optimization*, 24:51–60, 2002.

R.T. Haftka and Z. Gürdal. *Elements of Structural Optimization.* Kluwer, Dortrecht, 1992.

M.R. Hestenes. Multiplier and gradient methods. *Journal of Optimization Theory and Applications*, 4:303–320, 1969.

D.M. Himmelblau. *Applied Nonlinear Programming.* McGraw-Hill, New York, 1972.

W. Hock and K. Schittkowski. *Lecture Notes in Economics and Mathematical Systems. No 187: Test examples for nonlinear programming codes.* Springer-Verlag, Berlin, Heidelberg, New York., 1981.

N. Karmarkar. A new polynomial time algorithm for linear programming. *Combinatorica*, 4:373–395, 1984.

W. Karush. Minima of functions of several variables with inequalities as side conditions. Master's thesis, Department of Mathematics, University of Chicago, 1939.

J. Kiefer. Optimal sequential search and approximation methods under minimum regularity conditions. *SIAM Journal of Applied Mathematics*, 5:105–136, 1957.

H.W. Kuhn and A.W. Tucker. Nonlinear programming. In J. Neyman, editor, *Proceedings of the Second Berkeley Simposium on Mathematical Statistics and Probability.* University of California Press, 1951.

S. Lucidi and M. Piccioni. Random tunneling by means of acceptance-rejection sampling for global optimization. *Journal of Optimization Theory and Applications*, 62:255–277, 1989.

A.I. Manevich. Perfection of the conjugate directions method for unconstrained minimization problems. In *Third World Congress of Structural and Multidisciplinary Optimization*, volume 1, Buffalo, New York, 17–21 May 1999.

J.J. Moré and D. Thuente. Line search algorithms with guaranteed sufficient decrease. *ACM Transactions on Mathematical Software*, 20: 286–307, 1994.

J.A. Nelder and R. Mead. A simplex method for function minimization. *Computer Journal*, 7:308–313, 1965.

J. Nocedal and S.J. Wright. *Numerical Optimization.* Springer-Verlag, New York, 1999.

P.Y. Papalambros and D.J. Wilde. *Principles of Optimal Design, Modeling and Computation.* Cambridge University Press, Cambridge, 2000.

M.J.D. Powell. An efficient method for finding the minimum of a function of several variables without calculating derivatives. *Computer Journal*, 7:155–162, 1964.

S.S. Rao. *Engineering Optimization, Theory and Practice.* John Wiley, New York, 1996.

J.B. Rosen. The gradient projection method for nonlinear programming, Part I, Linear constraints. *SIAM Journal of Applied Mathematics*, 8: 181–217, 1960.

J.B. Rosen. The gradient projection method for nonlinear programming, Part II, Nonlinear constraints. *SIAM Journal of Applied Mathematics*, 9:514–553, 1961.

F. Schoen. Stochastic techniques for global optimization: A survey of recent advances. *Journal of Global Optimization*, 1:207–228, 1991.

J. A. Snyman. A gradient-only line search method for the conjugate gradient method applied to constrained optimization problems with severe noise in the objective function. Technical report, Department of Mechanical Engineering, University of Pretoria, Pretoria, South Africa, 2003. Also accepted for publication in the International Journal for Numerical Methods in Engineering, 2004.

J.A. Snyman. A new and dynamic method for unconstrained minimization. *Applied Mathematical Modelling*, 6:449–462, 1982.

J.A. Snyman. An improved version of the original leap-frog method for unconstrained minimization. *Applied Mathematical Modelling*, 7: 216–218, 1983.

J.A. Snyman. Unconstrained minimization by combining the dynamic and conjugate gradient methods. *Quaestiones Mathematicae*, 8:33–42, 1985.

J.A. Snyman. The LFOPC leap-frog algorithm for constrained optimization. *Computers and Mathematics with Applications*, 40:1085–1096, 2000.

J.A. Snyman and L.P. Fatti. A multi-start global minimization algorithm with dynamic search trajectories. *Journal of Optimization Theory and Applications*, 54:121–141, 1987.

J.A. Snyman and A.M. Hay. The spherical quadratic steepest descent method for unconstrained minimization with no explicit line searches. *Computers and Mathematics with Applications*, 42:169–178, 2001.

J.A. Snyman and A.M. Hay. The Dynamic-Q optimisation method: An alternative to SQP? *Computers and Mathematics with Applications*, 44:1589–1598, 2002.

J.A. Snyman and N. Stander. A new successive approximation method for optimal structural design. *AIAA*, 32:1310–1315, 1994.

J.A. Snyman and N. Stander. Feasible descent cone methods for inequality constrained optimization problems. *International Journal for Numerical Methods in Engineering*, 39:1341–1356, 1996.

J.A. Snyman, N. Stander, and W.J. Roux. A dynamic penalty function method for the solution of structural optimization problems. *Applied Mathematical Modelling*, 18:435–460, 1994.

N. Stander and J.A. Snyman. A new first order interior feasible direction method for structural optimization. *International Journal for Numerical Methods in Engineering*, 36:4009–4026, 1993.

K. Svanberg. A globally convergent version of MMA without line search. In G.I.N. Rozvany and N. Olhoff, editors, *Proceedings of the 1st World Conference on Structural and Multidisciplinary Optimization*, pages 9–16, Goslar, Germany, 1995. Pergamon Press.

K. Svanberg. The MMA for modeling and solving optimization problems. In *Proceedings of the 3rd World Conference on Structural and Multidisciplinary Optimization*, pages 301–302, Buffalo, New York, 17–21 May 1999.

H. Theil and C. van de Panne. Quadratic programming as an extension of conventional quadratic maximization. *Management Science*, 7:1–20, 1961.

A. Törn and A. Zilinskas. *Global Optimization: Lecture Notes in Computer Science*, volume 350. Springer-Verlag, Berlin, 1989.

G.N. Vanderplaats. *Numerical Optimization Techniques for Engineering Design.* Vanderplaats R & D, Colorado Springs, 1998.

G.R. Walsh. *Methods of Optimization.* John Wiley, London, 1975.

D.A. Wismer and R. Chattergy. *Introduction to Nonlinear Optimization. A Problem Solving Approach.* North-Holland, New York, 1978.

R. Zielinsky. A statistical estimate of the structure of multiextremal problems. *Math. Program.*, 21:348–356, 1981.

Index